Study Guide

Environmental Science Unit 1

for CAPE®

Alana Lancaster
Vindra Cassie
Philip Da Silva
Jillian Orford

OXFORD
UNIVERSITY PRESS

Great Clarendon Street, Oxford, OX2 6DP, United Kingdom

Oxford University Press is a department of the University of Oxford.
It furthers the University's objective of excellence in research, scholarship,
and education by publishing worldwide. Oxford is a registered trade mark of
Oxford University Press in the UK and in certain other countries

Text © Vindra Cassie, Phillip da Silva, Alana Lancaster, Jillian Orford 2014
Original illustrations © Oxford University Press 2015

The moral rights of the authors have been asserted

First published by Nelson Thornes Ltd in 2014
This edition published by Oxford University Press in 2015

British Library Cataloguing in Publication Data
Data available

978-1-4085-2348-3

14

Printed in Great Britain by CPI Group (UK) Ltd., Croydon CR0 4YY

Acknowledgements

Cover photography: Mark Lyndersay, Lyndersay Digital, Trinidad
www.lyndersaydigital.com
Page make-up: Fakenham Prepress Solutions, Norfolk
Illustrations: include artwork drawn by Fakenham Prepress Solution, Norfolk

The author and the publisher would like to thank the following for permission to reproduce material:

Text: p59, adapted from www.fao.org; p71, Annenberg Learner (www.learner.org) *The Habitable Planet,
Unit 5: Human Population Dynamics, Section 4: World Population Growth Through History*; p72 www.populationmatters.org
– Current population trends; p75, Copyright Guardian News & Media Ltd 2008; p76, www.iwhc.org, International
Women's Health Coalition, 'Child Marriage: Girls 14 and Younger at Risk'; p77, Los Angeles Times series Beyond 7
Billion Dream out of reach "Philippines birth control: Filipinos want it, priests don't" written by Kenneth R. Weiss,
published July 22, 2012; p82, John R. Weeks, 'How to Influence Fertility: The Experience So Far' (1990); p83,
The Jamaica Gleaner, 'Five-year Plan To Reduce Fertility Rate' by Anastasia Cunningham; p83, www.colby.edu; p84,
http://www.who.int/en/; p88, *Foreign Policy Magazine*.

Photographs: 1.3.2, Lee Dalton/Alamy; 1.3.3, travelib prime/Alamy; 1.7.1, Shutterstock; 1.7.2, iStockphoto;
1.7.3, iStockphoto; 1.11.2, Lee Dalton/Alamy; 1.18.1, iStockphoto; 1.21.1, Teddy/Alamy; 1.21.2, PetePhipp/Travelshots;
2.40.1, Shannon Fagan/Alamy; 2.44.1, Mike Greenslade/Alamy; 2.44.2, AFP/Getty Images; 2.48.1, Jacques Pavlovsky/
Sygma/Corbis; 2.50.1, Bob Krist/Corbis; 2.50.2, iStockphoto; 2.51.1, dbimages/Alamy; 3.53.1, Tony Rath's first
collection; 3.54.1, John Harper/Corbis; 3.54.2, iStockphoto; 3.54.3, Masa Ushioda/Alamy; 3.57.1, Image Source/
Alamy; 3.58.1, Eye Ubiquitous/Alamy; 3.59.1, Fotolia; 3.60.1, Shutterstock; 3.61.1, Fotolia; 3.62.1, Shutterstock; 3.62.2,
Peter Phipp/Travelshots.com/Alamy; 3.66.1, Shutterstock; 3.68.1, Shutterstock; 3.72.1, iStockphoto; 3.81.1, RANU
ABHELAKH/Reuters/Corbis; 3.82.1, Nigel Caitlin/Science Photo Library; 3.83.1, Adam Woolfitt/Corbis.

Contents

Introduction

This Study Guide has been developed exclusively with the Caribbean Examinations Council (CXC®) to be used as an additional resource by candidates, both in and out of school, following the Caribbean Advanced Proficiency Examination (CAPE®) programme.

It has been prepared by a team with expertise in the CAPE® syllabus, teaching and examination. The contents are designed to support learning by providing tools to help you achieve your best in CAPE® Environmental Science and the features included make it easier for you to master the key concepts and requirements of the syllabus. *Do remember to refer to your syllabus for full guidance on the course requirements and examination format!*

Inside this Study Guide is an interactive CD which includes electronic activities to assist you in developing good examination techniques:

- **On Your Marks** activities provide sample examination-style short answer and essay type questions, with example candidate answers and feedback from an examiner to show where answers could be improved. These activities will build your understanding, skill level and confidence in answering examination questions.

- **Test Yourself** activities are specifically designed to provide experience of multiple-choice examination questions and helpful feedback will refer you to sections inside the study guide so that you can revise problem areas.

This unique combination of focused syllabus content and interactive examination practice will provide you with invaluable support to help you reach your full potential in CAPE® Environmental Science.

Below are some points to consider when working through the exam-style questions at the end of each module.

- Pay attention to whether the question asks you to describe, comment on, assess, define, demonstrate, discuss, evaluate or explain, etc.

- The amount of detail required and the focus of your answer will depend on the instruction given in the question.

- The glossary provided in the syllabus is a good guide source of information as to what each rubric means.

- Approach all questions logically.

- The amount of marks indicated for a question is a good guide for you to know at least how many points the answer should have.

- Some questions do indicate how many points are required for the answer.

You can expect questions that ask you to distinguish between terms and concepts. Be prepared for these types of questions. Learning definitions and knowing the similarities and differences between terms, concepts and processes can help you prepare for such questions.

1.1   Introduction to ecology

Did you know?

Justus von Liebig (1840) stated that the single factor which is in short supply relative to demand is the critical factor that determines the distribution of a particular species. Victor Shelford further explained that each environmental factor has both maximum and minimum levels called tolerance limits beyond which a given species will fail to survive.

Ecological terms

Ecology is the study that deals with the relationships of organisms (species) to one another and their interactions with their physical surroundings. Ecology is therefore the study of relationship between organisms and their environment, both abiotic (physical or nonliving) and biotic (living).

Every living organism has limits to the conditions that it could survive, and the environmental factors must be within optimal levels for the organism to perform at its maximum. Among the environmental factors to which organisms must adapt are temperature, moisture levels, nutrient supply, soil and water chemistry and living space.

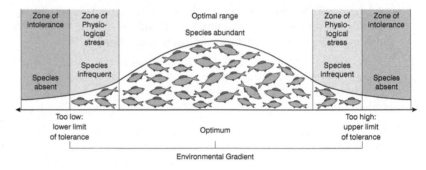

Figure 1.1.1 *The principle of tolerance limits*

There are times when the requirements and tolerances of species are good indicators of specific environmental characteristics. In fact the presence or absence of these environmental indicators can give information about the biological community and ecosystem.

Levels of biological organisation

Biology is studied at different levels of organisation. Biological organisation refers to a hierarchical system of classification in which each

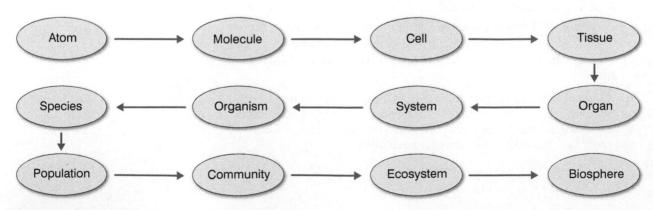

Figure 1.1.2 *Levels of organisation*

successive level is more complex than the lower level. The components that make up the different levels of biological organisation are either living or nonliving, are different from each other but are mutually dependent.

The species is the basic unit of biological classification and is a group of organisms that are genetically similar and interbreed with one another to produce live, fertile offspring.

A population is made up of all members of a species living in a given area at the same time. This is the fourth level of the biosphere. The addition or removal of a population can have serious consequences for an ecosystem. Indicator species are important in population studies because they give an indication of the health of the ecosystem, while keystone species can indicate the presence of certain other populations in the same ecosystem.

A biological community is the third level of organisation in the biosphere and is made up of all of the populations of organisms living and interacting in a particular area.

Communities share particular habitats and communities in a given location are limited to species that are capable of surviving the abiotic factors at that location. Biological communities are also limited by biological or biotic factors such as predators and available food resources.

An ecosystem comprises all of the populations in a given area (biological community) together with the nonliving components. The nonliving components include soil, water, light and nutrients. The living components may be either producers (plants) or consumers (animals).

Ecosystems may be divided according to interaction and energy transfer. Within an ecosystem, energy is consumed and matter is cycled between the different groups of organisms and between trophic levels (positions occupied in the food chain). Ecosystems vary in sizes and the organisms living in the ecosystem are adapted for life in such systems. Abiotic conditions of an ecosystem include physical and chemical factors. Examples of some of these factors are: sunlight, water, temperature, soil, wind, elevation and latitude.

Two major processes occur within ecosystems: flow of energy and cycling of nutrients.

An understanding of feeding relationships in ecosystems gives an indication of flow of energy and cycling of matter. Organisms in ecosystems are grouped according to how they feed. Autotrophs make their own food by photosynthesis and are called producers. Producers are photosynthetic organisms and include green plants, green algae and phytoplankton.

Consumers feed on other organisms and are classified based on the type of food that they eat. They can be classified as herbivores, carnivores, omnivores or detritivores. Consumers can also be described as primary consumers, secondary consumers or tertiary consumers. Omnivores feed on both plant and animal material. There are some organisms, called scavengers, that feed on dead organisms. Detritivores or detritus feeders feed on organic waste or pieces of dead organisms. Decomposers also feed on organic waste and dead organisms; however, they digest the materials outside of their bodies. Decomposers are important in recycling of nutrients.

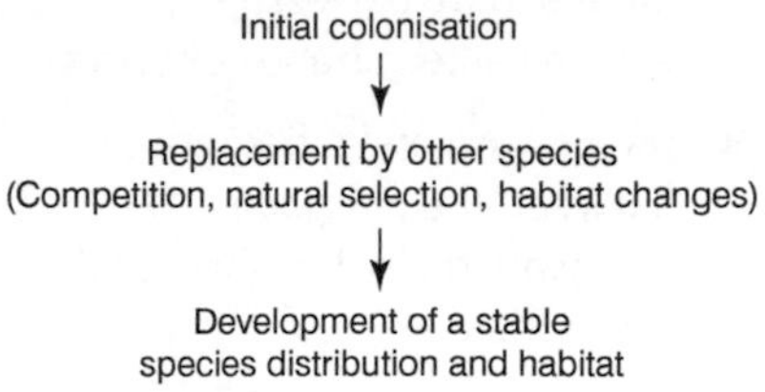

Initial colonisation

↓

Replacement by other species
(Competition, natural selection, habitat changes)

↓

Development of a stable
species distribution and habitat

Figure 1.1.3 *Sequence of ecological succession*

Ecological succession

In ecosystems populations fluctuate in response to environmental changes. These changes are often a part of the natural process called ecological succession. Ecological succession involves two types of succession: primary succession and secondary succession. Figure 1.1.3 represents a combination of these two types of succession.

Key points

- Ecosystems are made up of abiotic and biotic components.

- Studies of individuals are concerned with physiology, reproduction, development or behaviour.

- Studies of populations usually focus on the habitat and resource needs of individual species, their behaviour as a group, population growth and factors that limit their abundance and distribution.

- Studies of communities examine how populations of species interact with one another, in terms of predator–prey relationships and competition for common resources.

Learning outcomes

On completion of this section, you should be able to:

- differentiate between key ecological terms and concepts

- understand and differentiate between the biosphere, atmosphere, hydrosphere and lithosphere.

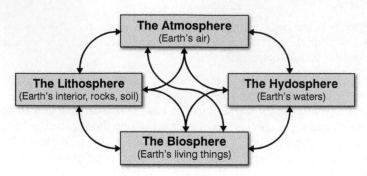

Figure 1.2.1 *The four interconnected parts of the Earth's system*

The biosphere, atmosphere, hydrosphere and lithosphere are four interconnected parts that form a system (Figure 1.2.1). Living things from the biosphere need air from the atmosphere and water from the hydrosphere to survive. They also need materials from the lithosphere such as minerals and related compounds. During these interactions living things modify the atmosphere and hydrosphere by adding and removing substances. Trees produce oxygen and release carbon dioxide and humans release pollutants into the atmosphere and hydrosphere. There are other examples of interconnections and interactions between the different spheres. You may wish to discuss some of these interconnections.

Atmosphere

The atmosphere is a layer of air that surrounds the planet Earth. Two gases make up the major part of the Earth's atmosphere: nitrogen (N_2), which comprises 78 per cent of the atmosphere, and oxygen (O_2), which accounts for 21 per cent. On the basis of temperature, the atmosphere is divided into four layers: the trophosphere, stratosphere, mesosphere and thermosphere.

Conduction, convection and radiation are responsible for transferring energy between the Earth's surface and the atmosphere. The physical and chemical structure of the atmosphere, the way in which the gases interact with solar energy, and the interactions between the atmosphere, land and oceans all combine to make the atmosphere an integral part of the biosphere.

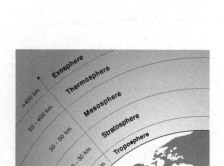

Figure 1.2.2 *Layers of the atmosphere*

Did you know?

The hydrosphere is important because:

- water is used by humans for different activities

- water is needed by living things for life processes and functions

- water is an important habitat for many organisms.

A planet's hydrosphere can be liquid, vapour or ice.

Hydrosphere

The hydrosphere is that part of the Earth that is composed of all of the water on or near the Earth. The hydrosphere includes water on the surface of the planet, underground and in the air. Water occupies almost 71 per cent of the surface area of the planet and includes the oceans, rivers, lakes and moisture in the air. Most of the Earth's water is present in the oceans, with the remainder existing as fresh water. The water in the hydrosphere is divided into oceanic waters (oceans and seas) and continental waters (water bodies on land, such as rivers, lakes and ice) and underground water.

The hydrosphere is always moving. Water in rivers flows to oceans and seas. The water in seas and oceans moves because of the action of the wind, which creates waves, currents and tides. In spite of this constant movement the quantity of water on the Earth remains fairly constant. As water moves it not only changes location but it changes form (liquid, vapour, and solid) as part of the hydrologic cycle.

Biosphere

The biosphere is the part of the Earth that supports all life. All living organisms are components of the biosphere. The biosphere is divided into regions called biomes and biomes are the largest of the five organisational levels. The other organisational levels are ecosystems, communities, populations and organisms. Biomes have no boundaries but instead have transition zones called ecotones, which contain a variety of plants and animals found in the adjoining biomes.

Biomes are defined on the basis of climate, geography and the species native to the region. The factors that determine climate include average temperature, amount of rainfall and humidity.

Aquatic biomes can be broken down into two regions: freshwater and marine. Freshwater biomes include ponds, lakes, streams, rivers and wetlands. Marine biomes include oceans, coral reefs and estuaries. Plants and animals living in freshwater biomes are adapted to life in areas of low salt content and cannot survive in areas of high salt content such as marine regions. Estuaries are areas where there is mixing of fresh water and marine water. These areas are described as being brackish. Estuaries support a diverse array of flora and fauna. Coral reefs are distributed in warm, shallow waters and may be found as barriers along continents or as fringing islands and atolls.

Lithosphere

The Earth is divided into three layers: the crust, the mantle and the core. The outermost layer is made up of the lightest material and the innermost layers are made up of the densest materials. The innermost layer is the core and it makes up approximately 33 per cent of the Earth's mass. The outermost layer is the crust and this layer is approximately 1 per cent of the Earth's mass. The middle layer is the mantle and it is approximately 66 per cent of the Earth's mass.

Two types of lithospheric crust are identified: continental crust and oceanic crust. Continental crust is thought to have a composition similar to granites while oceanic crust has a composition similar to basalt. Oceanic crust is denser than continental crust because basalt is denser than granite.

Key points

- The biosphere, atmosphere, hydrosphere and lithosphere are four interconnected parts that form a system.

- Conduction, convection and radiation are responsible for transferring energy between the Earth's surface and the atmosphere.

- The hydrosphere is that part of the Earth that is composed of all of the water on or near the Earth.

- All living organisms are components of the biosphere, which is divided into biomes.

- The Earth is divided into three layers: the crust, the mantle and the core.

Types of biome

The main types of biomes are: aquatic, desert, forest, grassland and tundra. One of the main reasons for classifying the biosphere into biomes is to highlight the important effect that physical geography has on communities of living organisms. Vegetation types are very useful when describing different biomes.

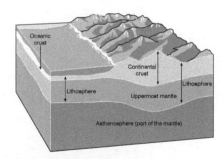

Figure 1.2.3 *The outer and upper layers of the Earth*

Physical layers

The Earth can be further divided into five main physical layers: the lithosphere, the asthenosphere, the mesosphere, the outer core and the inner core. The lithosphere is the outer solid part of the Earth and includes the oceanic and continental crusts and the upper mantle.

On completion of this section, you should be able to:

- differentiate between key ecological terms and concepts

- understand what habitats and ecotones are and differentiate between them.

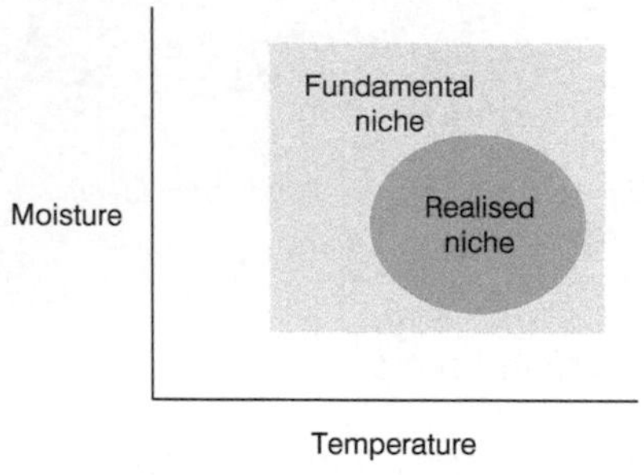

Figure 1.3.1 Relationship between fundamental niche and realised niche

Examples of biomes

Deserts, tropical rainforest, mangrove ecosystem, sea-grass beds, coral reefs, grasslands

Ecological terms

Habitat is the place where a particular organism lives and is characterised by a set of specific environmental conditions. The habitat of an organism is therefore the physical location where the organism is found.

The niche of an organism is the role that a species plays in a community. Niche and habitat are not the same thing. Many species share a habitat but every species has a unique niche.

The niche of an organism includes:

- its habitat

- its food resources

- its use of abiotic resources (light, CO_2, O_2, etc.)

- the way in which it is influenced by abiotic factors. Example: the maximum and minimum temperatures at which it can survive

- the way in which it interacts with other individuals of the same species and with individuals of other species.

If a species is to maintain its populations then the individuals must survive and reproduce. To do so, individuals of each species need to tolerate the physical environment, obtain energy and nutrients and avoid predators. Taken together, these factors which determine where the individuals of a species live and how abundant they are, are termed the ecological niche.

'Ecological niche' is a functional term that describes either the role played by a species in the community or the total set of environmental factors that determine species distribution. The niche describes how a species obtains food, the relationship it has with other species and the services it provides to the community.

A species has a fundamental niche and a realised niche. The fundamental niche is the set of favourable conditions that are determined by abiotic and biotic variables where the species can survive and successfully reproduce. The realised niche is where the species can persist given the presence of other species competing for the same resources.

The fundamental niche includes the total range of environmental factors that are suitable for the existence of the species without the influence of interspecific competition or predation. The realised niche is that part of the fundamental niche actually occupied by the species.

A biome is a type of ecosystem characterised by distinctive climate and soil conditions and a distinctive type of biological community that is adapted to the specific set of conditions. Biomes are characterised and defined by a set of complex interactions of the plants and animals with the climate, geology, soil, water and latitude of a given area.

Each biological community has spatial limits and sometimes the boundary between communities is sharp or gradual. An ecotone is

Figure 1.3.2 *Tropical rainforest biome*

Figure 1.3.3 *Coral reefs biome*

a transitional boundary between two ecological communities which contains characteristic species of each community and species peculiar to the ecotone. Generally the number of species and their population densities are greater in the ecotone than in adjoining communities. This increased density and diversity is because of the edge effect of two distinct communities. Ecotones are therefore important in facilitating the exchange of species and nutrients between communities.

Ecotones play an important role in conservation and ecological management, and as indicators of ecological change.

> **Activity**
> Make a table with the headings; biome, location, climate, soil, plants and animals. Complete the table with information about the various biomes and the features that define them.

Key points

- The habitat of an organism is the physical location where an organism is found.

- The niche of an organism is the role that a species plays in a community.

- The niche and habitat of an organism are not the same.

- While many species may share a habitat, every species has a unique niche.

- The niche of an organism includes its habitat, its food resources, its use of abiotic resources, the way in which it is influenced by abiotic factors and the way in which it interacts with other individuals of the same species and with individuals of other species.

- Biomes are characterised by a set of complex interactions of organisms with the climate, geology, soil, water and latitude of the given area.

- An ecotone is a transitional boundary between two ecological communities which contains characteristic species of each community as well as species peculiar to the ecotone.

On completion of this section, you should be able to:

- understand the meaning of key ecological terms and concepts

- understand and differentiate between abiotic and biotic components of the ecosystem

- discuss the concept of 'niche' and distinguish between fundamental and realised niche.

Figure 1.4.1 *Zonation*

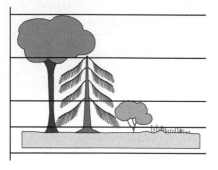

Figure 1.4.2 *Stratification*

Activity

- Which is greater, the ecological niche or the fundamental niche?

- Why is it not possible for two species to have identical ecological niches?

Ecological terms

There are many factors that contribute to the unique characteristics of an area. These factors may be classified as either **abiotic** factors or **biotic** factors. Biotic and abiotic factors combine to create the ecosystem. The ecosystem is a community of living and nonliving things considered as a unit.

The abiotic factors refer to the nonliving component of the environment. Examples of abiotic factors are temperature, pH, humidity, wind speed, air composition, amount of daylight, amount of light reaching certain areas, erosion rates, visibility, average rainfall, amount of space.

Biotic factors refer to living things, plants, animals, fungi, bacteria, protists (microorganisms) in the environment.

The biotic and abiotic factors in an environment are interrelated. The loss of a factor in the environment can affect the availability of other factors in the environment.

Abiotic and biotic factors determine structure and patterns in ecosystems

Abiotic and biotic factors vary temporally and spatially in ecosystems. Because of these variations, different spatial patterns exist in an ecosystem. Two spatial patterns that are recognised are **zonation** and **stratification**.

Zonation is the spatial pattern that occurs horizontally along the ground. Density and distribution of species vary along a horizontal gradient.

Stratification is the type of spatial pattern that occurs vertically and which is determined by the height of organisms. In a forest community, different species grow to different heights and this gives rise to stratification.

Succession is a temporal pattern that reflects change in species composition over time. Time is therefore important when considering **ecological succession**. Ecological succession goes through a sequence of stages starting with pioneer species. As time passes the diversity of species increases and eventually a climax community develops. It is important to note that the process does not simply end with the climax community, because the ecosystem is in continuous change – eventually giving way to secondary ecological succession.

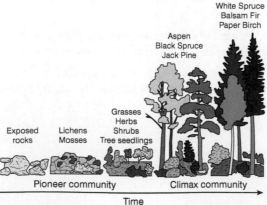

Figure 1.4.3 *Ecological succession*

Can two species have the same ecological niche?

Two species cannot share the same ecological niche, but individuals belonging to the same species may have different niches. **Resource partitioning** allows two species to coexist in a given location. However, competition between two species probably means that each species will have a more limited ecological niche than if it was living alone. The ecological niche when restricted by the presence of a competitor is the **realised niche**. The larger, potential niche that would occur without a competitor is the **fundamental niche**.

Each species has a fundamental niche and a realised niche. The fundamental niche is the set of favourable conditions that are determined by abiotic and biotic variables where the species can survive and successfully reproduce. The realised niche is where the species can persist given the presence of other species competing for the same resources.

The fundamental niche therefore includes the total range of environmental factors that are suitable for the existence of the species without the influence of interspecific competition or predation. The realised niche is that part of the fundamental niche actually occupied by the species (Figure 1.4.4).

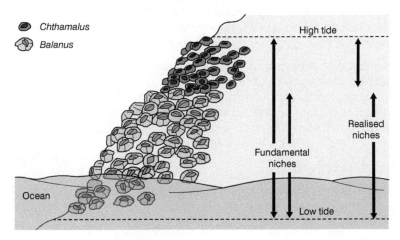

Figure 1.4.4 *Competition and niches*

The competitive exclusion principle and resource partitioning

The competitive exclusion principle states that two closely related species in competition for the same resources cannot coexist indefinitely and the species that is competitively inferior will eventually lose and be eliminated. This is particularly true if resources are limited. However, in nature species tend to develop mechanisms that promote coexistence rather than exclusion between closely related species.

One such mechanism in which species exist together as exceptions to the competitive exclusion principle is resource partitioning: the division of environmental resources between species. Very often behavioural differences tend to account for such ability to avoid competition that could totally exclude one of the competing species.

Examples of resource partitioning

Seabirds, cormorants (*Phalacrocorax carbo*) and shags (*Phalacrocorax aristotelis*) appear to have similar niches. They are often seen in the same coastal location searching for food. The two species look very similar and it is usually difficult to tell them apart. However, they are known to have different ecological niches.

Shags feed mainly on surface-swimming prey and choose sheltered coastal sites for breeding, such as crevices in rock gullies, ledges in the roofs of caves or among boulders on steep slopes. In contrast, cormorants feed mainly on bottom-feeding prey and prefer to breed on small rocky islands and broad cliff edges which are often more exposed. Cormorants usually need more space for their nests when compared to the nests of shags.

Key points

- Abiotic and biotic factors vary temporally and spatially in ecosystems and determine the structure and pattern of the ecosystem.

- Zonation occurs horizontally along the ground, whereas stratification occurs vertically and is determined by the height of organisms.

- Ecological succession is a temporal pattern that reflects change in species composition over time.

- Resource partitioning allows two species to coexist in a given location.

- In nature species tend to develop mechanisms that promote coexistence rather than exclusion between closely related species.

Liebig's law

According to Liebig's law of the Minimum, ecological events and their outcomes are often regulated by the availability of one or more factors which are in short supply. Although species are sensitive to all of the abiotic factors in their environment, the one factor that is in short supply, called a limiting factor, tends to regulate population size.

Limiting factors

Every organism has limits in terms of the extremes of abiotic and biotic factors that it can tolerate, and the distribution of organisms is also influenced by these limitations. Each organism has different requirements for growth and reproduction. Of all the abiotic and biotic factors that can affect populations only one, called a limiting factor, usually affects the population and limits its growth.

A limiting factor influences the distribution or population size of an organism or species. A limiting factor is any condition or factor that is outside of an organism's range of tolerance. Too much or too little of any abiotic factor can prevent the growth of a population even if all other factors are at or near the optimum range of tolerance. In ecosystems many factors interact and so it is often difficult to identify the one factor which limits population growth.

Plants and animals that succeed in occupying a particular niche tend to be those that adapt easily to the specific and at times unique environmental conditions at each location. Microclimatic conditions are among the most important factors that influence the successful establishment of plant and animal communities. Two of the more important climatic factors are sunlight and moisture.

The availability of water is important for the survival of almost all forms of animal life. Plants also require water for a number of life processes such as germination, growth and reproduction. In terms of plants the principle of limiting factors states that the maximum obtainable rate of photosynthesis is limited by whichever basic resource of plant growth is in least supply.

Range of tolerance

Each species can survive within a range of abiotic factors and every population has an optimal range of factors in which they thrive. Moisture levels, nutrients, soil and water conditions, temperature, living space and

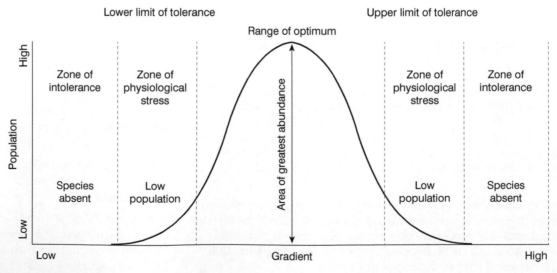

Figure 1.5.1 Range of tolerance

other environmental factors must be within appropriate levels for the life of organisms to persist. Each environmental factor has both maximum and minimum levels, called tolerance limits, beyond which a particular species cannot survive. Between the maximum and minimum levels is the range of tolerance. The point within this range where populations tend to flourish is called the optimum range of the species. Fewer and fewer organisms are found beyond this optimal range. Figure 1.5.1 shows different zones and population distribution at these different zones and ranges of factors.

Factors that influence life in aquatic ecosystems

There are several factors that limit life in aquatic ecosystems. Examples of some of these factors are phosphorus, pH, temperature, sunlight and dissolved oxygen.

Factor	Role of factor
Phosphorus	Phosphorus is a major limiting nutrient in freshwater ecosystems such as lakes, rivers and reservoirs. Naturally low levels of phosphorus keep populations of algae and other organisms in check. If phosphorus levels increase, for example by the introduction of sewage rich in phosphates, algal populations may explode and create algal blooms.
pH	Changes in pH affect organisms in aquatic environments as most organisms prefer a pH-neutral environment. A rise or fall in pH affects the rate of availability of minerals present in the water. This could affect the osmotic balance in organisms.
Temperature	The temperature of the water in aquatic ecosystems usually decreases with water depth, since less sunlight will penetrate the water at a greater depth. Most aquatic organisms have a limited range of tolerance to temperature changes. This is because temperatures are not likely to undergo major changes beneath the surface of the water. However, sudden temperature changes can affect the performance and survival of aquatic organisms because they have a specific range of temperature to which they are adapted. Temperature changes also affect enzyme activity in physiological and metabolic reactions. Temperature may also have a direct effect on organisms by affecting diurnal or temperature-related activities.
Sunlight	Sunlight can only penetrate water up to certain depths below the surface. Sunlight influences photosynthesis in primary producers and, ultimately, productivity of the ecosystem. Producer organisms need sunlight to produce oxygen and other required substances that will sustain consumers. Primary production will only occur in the zone where sunlight can penetrate.
Dissolved oxygen (DO)	Dissolved oxygen concentrations are influenced by water temperatures. Oxygen level, like temperature, decreases with depth. When dissolved oxygen levels fall below a certain point many consumer organisms, such as fish and zooplankton, will die. Hence dissolved oxygen and water temperature are very important limiting factors in aquatic ecosystems. Oxygen is essential for respiration in organisms and aquatic organisms obtain their supply from oxygen that is dissolved in the water. If DO levels decrease then organisms will experience reduced ability to engage in aerobic respiration. This will eventually affect energy output and metabolic activities. Anoxic conditions and anaerobic conditions will then develop.

Key points

- A limiting factor influences the distribution or population size of an organism or species.
- Each species can survive within a range of abiotic factors and every population has an optimal range of factors in which they thrive.
- Certain factors limit life in aquatic ecosystems.

Learning outcomes

On completion of this section, you should be able to:

- distinguish between the different types of competition in ecosystems

- explain predator–prey interactions

- discuss types of interactions between organisms in the environment.

Interaction and competition for resources

In ecosystems organisms are continually interacting with each other. Interactions are either between members of the same species or between members of different species. Some ways in which interactions are manifested include:

- exploitation of resources, which leads to depletion of resources

- pre-emptive and territorial interaction, which concerns space utilisation

- chemical production of toxins

- encounter involving transient interactions directly over a specific resource.

Interactions among organisms may be classified in terms of whether the interaction:

- is beneficial to individuals of all interacting species

- benefits individuals of one species but is harmful to those of another species

- is harmful to individuals of both species

- benefits individuals of one species but has no significant effect on individuals of the other interacting species.

Interactions among individuals of different species are called relationships.

Competition

Ecological interactions also manifest as different forms of competition. Competition can be defined as the use of a resource by one individual that reduces the availability of that resource for another individual. Competition occurs between individuals for different resources: food, shelter, mates, space, nutrients.

It is important to understand that the ecological definition of competition does not emphasise physical altercation because ecological competition can take place between individuals that never see or physically combat each other. Competition in ecosystems results whenever one individual or species reduces the availability of limited resources for another individual or species.

There are two types of competitive interactions, regardless of whether the interaction is intraspecific or interspecific. In exploitation competition, while all individuals have equal access to a particular resource, individuals differ in their ability to exploit the given resource. In interference competition, some individuals limit the access of others to a given resource.

Exploitative competition occurs when consumption of a limiting resource by one species makes that resource unavailable for consumption by another. Some examples of resources that organisms compete for include food, water, space, mates, soil nutrients and light.

Exploitative competition is a type of indirect ecological interaction and is a common mechanism of competition in nature. Exploitative

Did you know?

Examples of exploitative competition include:

- jaguars competing for food in the Guyana savannahs

- two species of barnacles competing for space on a rocky shoreline

- two species of savannah plants competing for light, soil moisture and soil nutrients.

competition may occur within species (intraspecific competition) or between species (interspecific competition). Exploitative competition within species can and does play a role in limiting population sizes. Exploitation competition between species can also influence the population sizes of competing species. Competitive exclusion, due to exploitative competition, can influence the number of species that can coexist in a community.

Intraspecific competition

Intraspecific competition is competition among members of the same species. Generally intraspecific competition is stronger and more intense than interspecific completion. This is because individuals of the same species have very similar resource requirements and if these resources are limiting, then individuals will compete for access to these resources. Intraspecific competition can decrease the reproduction and survival of individuals in a population as the carrying capacity of the population is approached.

Interspecific competition

Interspecific competition is competition between members of different species. Individuals of one species experience reduced fecundity, survivorship or growth as a result of resource exploitation or interference by individuals of another species.

Interspecific competition is usually weaker because two species never use exactly the same resources since they do not have the same ecological niche. When measuring or evaluating the impact of interspecific competition on population growth one must consider the amount of resources used and the overlap in the set of resources used.

Competition leads to exclusion and resource partitioning

Competitive exclusion suggests that complete competitors cannot exist indefinitely. Coexistence is probable when the niches of competitors do not overlap and differences in adaptive traits usually give some species a competitive advantage. Resource partitioning is important when similar species share the same resource in different ways. Resource partitioning may occur either when ecological differences between established and competing populations increase through natural selection or when species that are dissimilar succeed in joining an existing community.

Predator–prey interaction

A predator is any organism that gains its nourishment from killing and then eating other animals. Predator species need to be efficient hunters if they are to catch enough prey to survive. Prey, on the other hand, must be efficient at escaping from their predators if they are to survive and continue the perpetuation of the species. Growth in the prey population in an ecosystem results in an increase in the predator population in response to an increase in the potential food supply. As the population of predator increases the population of the prey population will reduce to an extent where it can no longer sustain the predator population. The predator population is therefore limited by its food supply and the prey population is determined by the number of individuals killed by the predator. Figure 1.6.1 shows this dynamic between population numbers of predator and prey.

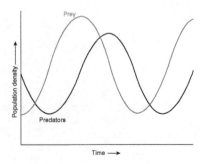

Figure 1.6.1 *Predator–prey population interaction*

Key points

- Competition is an interaction between individuals that results from a shared requirement for a resource that is in limited supply.

- Fecundity refers to the state of being fertile and capable of producing offspring.

- Predators differ from parasites in that they do not kill their host to obtain food.

On completion of this section, you should be able to:

- discuss types of interactions between organisms in the environment
- explain symbiotic relationships
- distinguish between different types of symbiotic relationships.

Symbiosis

Symbiosis is a relationship between individuals of two or more different species. Symbiosis is always 'inter-specific' because it occurs only between different species. Sometimes symbiotic relationships benefit both species, sometimes only one species benefits at the expense of the other species and in some instances neither species benefits.

The basis of symbiosis is that there is always competition for food and space among different organisms and species. Thus forming stable relationships with another species other than predator–prey relationships allows two different species to share the same space and or food supply.

Symbiotic relationships may be divided into three main categories:

Mutualism	Both species benefit from the relationship.
Commensalism	One species benefits and the other is not affected.
Parasitism	One species benefits and the other is harmed in the process.

Some symbiotic relationships are obligate relationships in that the organisms depend on each other for their survival. There are some symbiotic relationships that are deemed facultative, which means that they are not absolutely necessary for the survival of either species involved in the relationship.

Mutualism

A mutualistic relationship occurs when two organisms of different species interact and each benefits from the relationship.

Cattle egret and cattle	The cattle egret eats ticks, which are ectoparasites on the cattle. The egret gets food and the cattle get pests removed.
Honey bee and flowers	Honey bees fly from flower to flower to pollinate the flowers while gathering nectar, which they use to make food. The bees benefit by getting food while the flower benefits by getting pollinated.
Intestinal bacteria and humans and some other animals	A certain type of bacteria lives in the intestines of humans and other animals. The bacteria help in the partial digestion of some of the food that humans cannot digest initially.

Commensalism

This type of symbiosis usually takes place between one species that is either vulnerable to predation or with an inefficient means of locomotion and another species with an effective defence system.

Figure 1.7.1 *The cattle egret eats ticks on the cattle*

Figure 1.7.2 *Honey bee and flower*

| Gobies and sea urchins | Goby species live among the spines of the toxic sea urchins and gain protection from their host. |
| Orchids attached to trees | Orchids attach to trees to get water that drips from the tops of trees. The tree is neither harmed by, nor benefits from, this association. |

Figure 1.7.3 *Orchid attached to a tree*

Parasitism

In a parasitic relationship one of the organisms in the relationship lives off the other organism, the host, harming it and sometimes causing death. The parasite lives on (ectoparasite) or in (endoparasite) the body of the host. Although parasites may harm their hosts, it is in the parasite's best interests not to kill the host since it is dependent on the host's body and functions for its survival. Parasitism occurs in both animals and plants. Some examples of parasitic relationships are:

Aphids and plants	Aphids are insects that feed on sap from the plants on which they live.
Parasitic fungi and plants and animals	Wheat rust is caused by a parasitic fungus. The downy mildew fungus attacks fruits and vegetables.
Fungus and animals	Lumpy jaw disease is caused by a fungus that affects the jaws of cattle and pigs.
Insects and animals	Ticks and fleas affect dogs and other animals. Fleas bite the skins, suck blood and cause severe itching in the host. Fleas in turn get food and a warm home.

Key points

- Symbiosis is an ecological relationship between two organisms of different species living closely together and forming some type of feeding relationship between them.

- Symbioses are a dynamic way in which different species interact with each other.

- These interactions are either positive or negative depending on the nature and extent of involvement of organisms with each other.

The role of carbon and carbon dioxide

Carbon is an essential element in all living things. Carbon is also a part of oceans, the atmosphere and even rocks in the earth. Carbon serves two main functions in the lives of organisms: it is a structural component of organic molecules and it is an important element in chemical energy storage.

In the atmosphere carbon exists mainly as carbon dioxide. Carbon dioxide is used by plants in photosynthesis and carbon becomes a part of the plant.

Carbon dioxide is a greenhouse gas and it therefore traps heat in the atmosphere. However, while it is important in preventing the Earth from being frozen in its absence, anthropogenic activity has resulted in a large buildup of this greenhouse gas in the atmosphere.

When plants die and become buried over millions of years they may turn into fossil fuels. Humans burn fossil fuels and most of the carbon is released into the atmosphere as carbon dioxide.

The carbon cycle

The element carbon moves or is cycled between the living (biotic) and nonliving (abiotic) environments. A number of different biochemical and physical processes are responsible for driving the carbon cycle. Gaseous carbon is fixed in the process of photosynthesis and returned to the atmosphere in respiration.

Carbon dioxide in the atmosphere is the most accessible source of carbon. Aerobic respiration is responsible for returning carbon to the atmosphere in the form of carbon dioxide:

$$C_6H_{12}O_6 \text{ (aq)} + 6O_2 \text{ (g)} = 6CO_2 \text{ (g)} + 6H_2O \text{ (l)}$$

Combustion is a process that releases carbon dioxide into the atmosphere when fossil fuels such as coal, wood and hydrocarbons are burnt. Combustion can be due to natural activities (forest fires during *El Nino* events) or as a result of human activities (use of fuel for transport or for cooking).

Carbon fixation occurs during the process of photosynthesis when plants remove carbon dioxide from the atmosphere to make carbohydrates. Sunlight provides the energy for this reaction:

$$6CO_2 \text{ (g)} + 6H_2O \text{ (l)} = C_6H_{12}O_6 \text{ (aq)} + 6O_2 \text{ (g)}$$

Carbon dioxide is very soluble and large amounts are removed from the atmosphere when it dissolves in the oceans:

$$CO_2 \text{ (g)} + H_2O \text{ (l)} = HCO_3^- + H^+ \text{ (aq)}$$

Weathering and precipitation account for a small portion of the turnover of carbon in the biosphere.

Sedimentation is also responsible for the removal of some carbon from the carbon reservoirs. Calcareous rocks such as limestone and chalk are formed from the skeletal remains of microscopic organisms which previously combined carbon dioxide with calcium to form calcium carbonate ($CaCO_3$).

The role of organisms in the carbon cycle

Different organisms play different roles in the carbon cycle. Three examples of organisms and their roles are:

Termites digest cellulose in plant material, breaking it down and releasing carbon dioxide in the process back to the ecosystem.

Fungi are decomposers that break down dead material for use as food, converting it into fungal biomass and thus making it available for return to the food chain. During the decomposition process some of the carbon in the complex organic molecules may be released back into the atmosphere as carbon dioxide, or sometimes as methane, which is eventually oxidised to produce carbon dioxide. The decomposition process may also result in the breakdown of complex organic molecules in a way which converts nitrogen, phosphorus and potassium into the simple inorganic molecules which plants can reuse.

Dung beetles make it possible for carbon to re-enter the food chain when the larvae feed on the cow manure and make it available to decomposers. Dung beetles are nature's way of recycling carbon and minerals back into the soil to be broken down further into humus for plants. Dung beetles feed on faeces in both their larval and adult forms. Dung pats contain carbon and valuable soil nutrients and are a food source for soil microflora (fungi, bacteria and actinomycetes), protozoa, and earthworms.

Humans and the global carbon cycle

Anthropogenic activity affects the carbon cycle and the amount of carbon stored in carbon sinks. Humans deplete fossil fuel reserves while increasing the amount of carbon dioxide in the atmosphere through the burning of fossil fuels. This increases global warming and the level of air pollution. Reduction in the use of fossil fuels through the use of alternative sources of energy that are less polluting (solar energy, wind energy and hydropower) can help reduce these negative impacts.

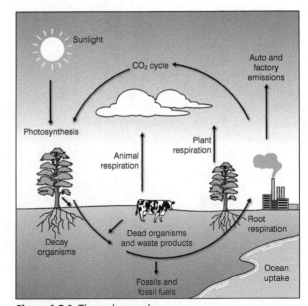

Figure 1.8.1 *The carbon cycle*

Key points

- The carbon cycle represents the circulation and reutilisation of carbon atoms in nature.

- Because of the carbon cycle the level of carbon dioxide in the atmosphere is relatively constant.

- Respiration and photosynthesis are two main processes that help to maintain the balance in the carbon cycle.

- Natural fires and human-created combustion can upset the balance in the carbon cycle.

Did you know?

Although nitrogen is very abundant in the atmosphere as dinitrogen gas (N_2), it is largely inaccessible in this form to most organisms. Nitrogen is therefore considered a scarce resource, and this scarcity often limits primary productivity in many ecosystems.

The role of nitrogen

Organisms cannot exist without the nitrogen-containing organic molecules amino acids, peptides, polypeptides and proteins. Green plants (producers) incorporate elemental nitrogen from the environment and use it to build their own protein molecules, which are eventually eaten by consumers.

Nitrogen becomes available to primary producers, such as green plants, only when it is converted from dinitrogen gas (N_2) into ammonia (NH_3). In addition to N_2 and NH_3, nitrogen exists in many different forms, including both inorganic (e.g. ammonia, nitrate) and organic (e.g. amino and nucleic acids) forms. Nitrogen undergoes many different transformation processes in the ecosystem. The major transformation processes for nitrogen are nitrogen fixation, nitrification, denitrification and ammonification (Figure 1.9.1). The transformation of nitrogen into its many oxidation states is important to productivity in the biosphere and is highly dependent on the activities of microorganisms.

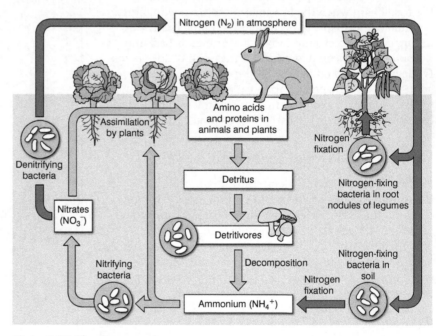

Figure 1.9.1 The nitrogen cycle

The nitrogen cycle

Nitrogen fixation: this process reduces nitrogen to ammonium ions or ammonia. This multistage process is catalysed by the enzyme nitrogenase. Free-living nitrogen-fixing bacteria *Azotobacter* as well as symbiotic bacteria such as *Rhizobium* and cyanobacteria (blue-green algae) such as *Nostoc* carry out this process. Some nitrogen-fixing bacteria are found in the root nodules of legumes.

Lightning may also fix atmospheric nitrogen: High energy discharge during lightning may combine nitrogen and oxygen. The nitrogen oxides that are formed may dissolve in rainwater to form weakly acidic solutions.

$$N_2 + 2O_2 \rightarrow 2NO_2 + H_2O \rightarrow 2HNO_3$$

$$N_2 + O_2 \rightarrow 2NO + H_2O \rightarrow 2HNO_2$$

Figure 1.9.1 Nitrogen oxidised by lightning dissolves in water

Nitrification: This is a series of oxidations which convert ammonium compounds to nitrate with the subsequent release of energy for the bacteria involved. Nitrite-forming bacteria combine ammonia with oxygen to form nitrites (NO_2^-), and another group of bacteria convert the nitrites to nitrates (NO_3^-).

$$NH_4^+ \xrightarrow{\textit{Nitrosomonas}} NO_2^- + \text{energy}$$

$$NO_2^- \xrightarrow{\textit{Nitrobacter}} NO_3^- + \text{energy}$$

The nitrates produced represent the form of nitrogen that is easily absorbed by the roots of plants. After nitrates are absorbed they are converted to ammonium (NH_4^+), then amino acids, peptides, polypeptides and proteins.

Denitrification: this is the process in which denitrifying bacteria break down nitrates into nitrogen (N_2) and nitrous oxide (N_2O), which are returned to the atmosphere:

$$2NO_3^- \xrightarrow{\textit{Pseudomonas}} N_2 + 3O_2$$

$$S + 2NO_3^- \xrightarrow{\textit{Thiobaccilus}} N_2 + SO_4^{2-} + O_2$$

Denitrification is a prevalent process in waterlogged soils which have a large quantity of decomposable organic matter and low oxygen availability.

Humans and the nitrogen cycle

Humans impact the nitrogen cycle by either adding or removing from the amount of nitrogen that enters the nitrogen cycle. The use of fertilisers adds nitrogen compounds to the soil. This type of anthropogenic activity accounts for a large amount of nitrogen fixation annually. Cultivation of leguminous crops increases the amount of nitrogen-fixing microorganisms in the soil.

Burning of fossil fuels increases the concentration of nitrous oxide (N_2O) and ammonia (NH_3) in the atmosphere while also increasing the concentration of other nitrogen oxides, such as nitric oxide (NO).

Humans have increased aquatic pollution through an increase in nitrate runoff in aquatic systems. Such increases result in eutrophication and further pollution, which affects some animal and plant species.

Key points

- The nitrogen cycle represents the circulation and reutilisation of nitrogen in nature.
- The nitrogen cycle is very dependent on the activities of microorganisms.
- Nitrogen fixation by nitrogen-fixing bacteria and lightning is responsible for 'fixing' atmospheric nitrogen in a form that can be used by plants.
- Nitrification results in the formation of nitrates, which are easily absorbed by plants.
- Denitrification represents a loss of nitrogen from the ecosystem and may occur under aerobic and anaerobic conditions.
- Anthropogenic activities also add to greenhouse gases, acid rain and pollution.

The important role of phosphorus

The energy-rich phosphorus-containing compounds phosphates and adenosine triphosphate (ATP) are important in energy transfer reactions at the cellular level. Phosphate is incorporated into the backbone of the nucleic acids deoxyribonucleic acid (DNA) and ribonucleic acid (RNA), which are important in genetics. Phosphorus is important to vertebrates; in the body of humans approximately 80% of phosphorus is present in teeth and bones.

Phosphorus is a limiting factor in aquatic ecosystems. The amount of available phosphorus influences the productivity of the ecosystem. High concentrations of phosphorus in aquatic ecosystems promote lush growth of algae and vegetation. Death and subsequent decomposition of the short-lived algae by oxygen-consuming bacteria causes pollution of the water body.

The phosphorus cycle

The phosphorus cycle is different from other biogeochemical cycles in that it is a sedimentary biogeochemical cycle because it does not have a gaseous phase. Phosphorus is transported in aqueous form.

The largest reservoir of phosphorus is in sedimentary rocks, where this cycle begins. When it rains, phosphates are removed from the rocks through the process of weathering and find their way into both soil and water. Phosphates are less soluble than nitrates and phosphates usually move to water bodies by the process of erosion and not by leaching. By the process of weathering and erosion phosphates enter rivers and streams, which transport them to the ocean. Once in the ocean the phosphorus accumulates on continental shelves in the form of insoluble deposits which are exposed on land after millions of years.

Phosphate ions are taken up by plants from the soil. These ions move from plants to animals when the plants are eaten by herbivores and when carnivores feed on the herbivores. The phosphate that is absorbed by animal tissues through consumption is eventually returned to the soil when animals get rid of waste products of metabolism. Phosphate is also returned to the environment when bacterial decomposition of organic matter takes place.

The same processes involved in the cycling of phosphorus in terrestrial ecosystems take place in aquatic ecosystems. Deep ocean sediments are important phosphorus sinks. Phosphorus usually binds to soil particles and enters waterways through sediment runoff. Phosphates are washed into waterways and are utilised by aquatic vegetation and algae. These phosphates may come from fertiliser runoff, sewage discharges, natural deposits and industrial wastes. Phosphates tend to settle on the bottom of oceans and lakes and other water bodies. When these sediments are stirred up, phosphates then re-enter the phosphorus cycle.

Phosphates are beneficial to plants but an excessive quantity is a pollutant in aquatic ecosystems. Excess phosphates lead to eutrophication and stimulate excessive growth of algae, plankton aquatic

Did you know?

Phosphorus also enters the environment from rocks or other deposits such as fossilised bone or bird droppings called guano.

Did you know?

Phosphorus re-enters the cycle when phosphorus-containing organisms are eaten, but mainly by being incorporated into ocean sediments that over geological time are returned to the terrestrial environment through mountain-building (orogenesis).

plants. This has severe implications for the amount of dissolved oxygen in the water body, resulting in fishes and other aquatic organisms suffering from limited supply of oxygen. Such excess growth reduces sunlight penetration, thus limiting the amount of light reaching benthic-dwelling species.

Human activities that influence the phosphorus cycle:

- Commercial synthetic fertilisers.
- Mining of deposits of calcium phosphate.
- The use of large amounts of sulphuric acid to convert phosphate rock into superphosphate fertiliser.
- Cutting tropical rainforests.
- Agriculture runoff results in large quantities of phosphates entering aquatic systems, causing eutrophication.
- Some detergents contribute to large amounts of phosphates reaching waterways.

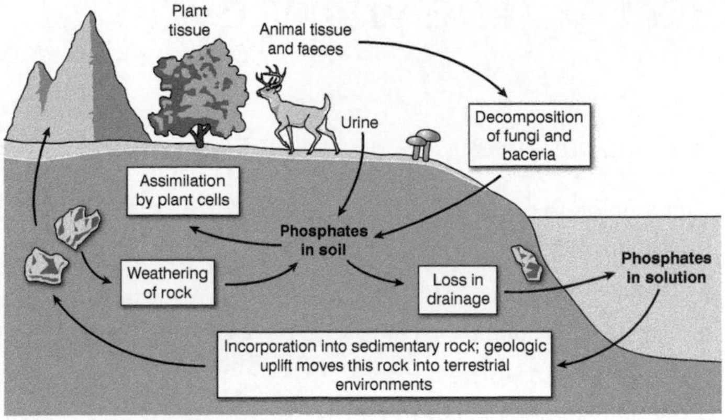

Figure 1.10.1 *The phosphorus cycle*

Why is eutrophication a problem?

Nutrient enrichment of aquatic ecosystems (eutrophication) poses serious problems, directly or indirectly, for water bodies and the organisms that need such habitats.

When the aquatic algae die and decay, dissolved oxygen in the water is used up and the water may become anoxic (no dissolved oxygen in the water), leading to the formation of foul-smelling odours associated with the presence of hydrogen sulphide (H_2S) and ammonia (NH_3) and thioalcohol (RSH). Thioalcohols are alcohols in which the functional group has sulphur instead of oxygen.

High concentrations of ammonia can result in reduced growth, increased susceptibility to disease and can be lethal to fish. Two forms of ammonia exist: un-ionised ammonia (NH_3) and ionised ammonia, also known as ammonium ion (NH_4^+). The ratio of un-ionised to ionised ammonia depends on the pH and temperature of the water. Un-ionised ammonia (NH_3) is extremely toxic to fish and is the predominant form of ammonia when water pH is high. Ionised ammonia (NH_4^+) is nontoxic, except at extremely high levels, and is the predominant form present when pH of water is low.

Some key effects of eutrophication are:

- Anoxia leads to fish-kills and release of unpleasant gases and odours.
- Algal blooms and uncontrolled growth of aquatic vegetation.
- Increased turbidity levels of the water.
- Decreased productivity/fish yields due to reduced oxygen levels in the water.
- Oxygen depletion of deeper layers in the aquatic ecosystem.
- Production of toxic substances by increased growth of certain algal species.
- Deterioration of recreational potential of water bodies.
- Changes in species composition in the water body.

Key points

- The phosphorus cycle represents the circulation and reutilisation of phosphorus in nature.
- The phosphorus cycle is a sedimentary cycle and therefore does not have a gaseous phase.
- It therefore takes a very long time for phosphorus atoms to pass through the cycle.
- Phosphorus is a limiting factor in the growth of algae and plants.

On completion of this section, you should be able to:

- outline the water cycle
- explain the significance of the water cycle
- write equations to represent the different key processes in the water cycle.

The role of water

Water covers about 70 per cent of the Earth's surface. As a basic compound in nature, water needs to be replenished and purified so that the functions that it provides and supports can be realised. The water cycle sees water moving through three phases (solid, liquid and gas) over the four spheres (atmosphere, lithosphere, hydrosphere and biosphere).

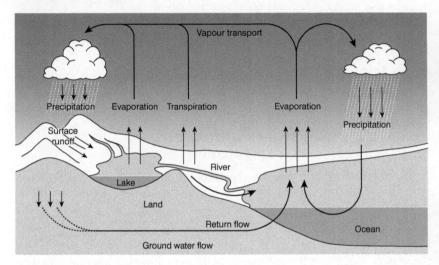

Figure 1.11.1 *The water cycle*

The water cycle

The water cycle is also known as the hydrologic cycle and describes the different processes by which the various forms of water move about the planet. There are a number of basic processes that drive the water cycle: evaporation, condensation, precipitation and runoff.

Evaporation: this process occurs when, with the help of the sun's rays, water transforms from liquid to gas. Evaporation includes evapotranspiration from plants. The sun is the driving force of the water cycle. It provides the energy to heat up the water, which evaporates and rises into the atmosphere.

Condensation: occurs when vapour rises into the atmosphere, forming clouds and fog. As the water vapour rises higher in the atmosphere, it cools due to a decrease in temperature. On cooling condensation takes place, resulting in the formation of tiny droplets of water. Once clouds are formed, advection takes place. Without advection the cycle would cease since water would evaporate and precipitate in the same place.

Precipitation: When the cloud can no longer hold any more condensed water vapour this water returns to the Earth as rain, snow, sleet or hail, known collectively as precipitation. While most of this returns to the Earth, some is intercepted by plants and evaporates back into the atmosphere instead of falling to the ground.

Did you know?

Advection in the water cycle is the movement of water through the atmosphere.

Runoff: This is the term used to describe the water which falls back to the surface of the Earth and which either stays on the surface of the Earth or enters different water bodies. Runoff includes both surface runoff when water travels over land, and channel runoff when water gets into streams and rivers. As the water moves it can drain underground, evaporate into the air, be stored in lakes or reservoirs or harvested for human use.

Infiltration: Sometimes referred to as percolation, this is the process in which water on the surface of the Earth seeps down into the ground to form aquifers.

Transpiration: This is the process by which plants absorb water from the soil. This water moves from the roots, up the stem and to the leaves. Water also evaporates off the leaves and into the atmosphere by a process called evapotranspiration.

Figure 1.11.2 *The Kaieteur falls in Guyana*

Key points

- The water cycle represents the circulation and reutilisation of water in nature.
- As a result of evaporation, condensation and precipitation, water travels from the surface, goes into the atmosphere and returns to the Earth.
- Water condenses into droplets in the presence of small dust particles around which the droplet can form.
- As water infiltrates through the soil and rocks, many of the impurities in the water are filtered out, resulting in the cleansing of the water.
- Surface runoff is important because it allows much of the water to return to the ocean, where a lot of evaporation takes place.
- The water cycle influences climate and weather patterns and therefore changes as global climate changes.

Learning outcomes

On completion of this section, you should be able to:

- distinguish between food chains and food webs

- explain trophic levels in ecosystems

- explain how energy flows in ecosystems.

Energy flow in ecosystems

The energy from the sun is the fundamental source of energy that sustains most terrestrial and aquatic ecosystems. This energy exhibits a unidirectional flow; from producers (autotrophs) which harness the light energy from the sun for photosynthesis to consumers that feed on the autotrophs and each other. This unidirectional flow of energy is a direct consequence of the law of thermodynamics, which refers to the fact that energy cannot be created or destroyed but is transformed from one form to another.

In ecosystems there are many roles in relation to food chains and food webs that are played by producers, consumers and decomposers. The pathway of energy flow in a food chain web begins with producers. The energy in the organic matter that is synthesised by producers is transferred to consumers when the producers are eaten. The transfer of energy which takes place in ecosystems can be shown in a number of ways, such as food chains, food webs and ecological pyramids.

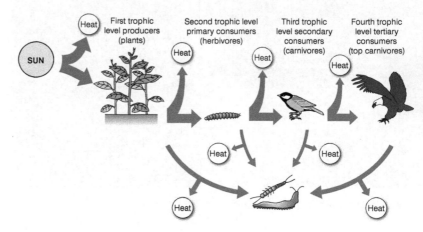

Figure 1.12.1 *Energy flow through different trophic levels*

Food chains and food webs

The transfer of food energy from producers through a series of consumers forms a 'food chain'. The arrows in a food chain represent the direction of flow of energy, with the head of the arrow always pointing away from the organism that is being eaten. In a food chain each organism represents a trophic level. Food chains do not have unlimited links or trophic levels because at each trophic level a high percentage of the useful energy (90%) consumed as food is converted to heat and is lost from the food chain. Energy is also lost when all of the preceding organism is not eaten and when the organism gets rid of faeces. Hence only 10 per cent of the energy at each trophic level is available for transfer to the next higher trophic level. Hence shorter food chains provide more energy to consumers. Some omnivores occupy more than one trophic level and some organisms occupy different trophic positions in different food chains.

In nature food chain relationships are not isolated because most organisms have a wide range of food sources, and therefore belong to

Did you know?

Food chains are of two types: grazing food chains and detritus food chains. Grazing food chains start with photosynthesising organisms (autotrophs or producers), followed by grazing herbivores and then carnivores. Detritus food chains start with dead organic matter followed by microorganisms and then organisms that feed on detritus and their predators.

more than one trophic level. The many food chains in an ecosystem are usually interconnected to form food webs. Food webs are therefore complex networks of interconnected food chains at different trophic levels. Food webs allow an organism to obtain its food from more than one type of organism of a lower trophic level. Food webs do not, however, show how much energy flows from one trophic level to the next in an ecosystem.

Bioaccumulation and biomagnification

Bioaccumulation and biomagnification allow us to understand how pollutants traverse different trophic levels. Bioaccumulation refers to the way pollutants enter a food chain. This process is responsible for the increase in concentration of a pollutant from the environment to the first organism in a food chain. Biomagnification refers to the tendency of pollutants to increase in concentration as they traverse one trophic level to the next higher trophic level. Biomagnification is responsible for increases in the concentration of a pollutant from lower to higher trophic levels.

These two processes are important because they are effectively responsible for enabling small concentrations of pollutants or toxic substances in the environment to find their way into organisms through the food chain, eventually resulting in problems for the organisms.

It is more energy efficient for humans to eat corn rather than beef.

A good way to analyse feeding relationships of a community is to analyse trophic levels. It is more energy efficient for humans to eat corn rather than beef because corn has more stored potential energy than beef.

Plants convert solar energy to chemical energy, which is stored in the grains of corn. As the cow eats the corn, there will be a loss of some of this energy so less energy will be available for conversion to biomass in the cow. Only 10 per cent of the energy stored in the corn is transferred to the cow. Ninety per cent of the energy is lost through transfer between trophic levels and hence much less energy will be available for humans.

The longer the food chain the less energy is available at the end of the chain. Thus, when humans eat corn directly, more of the stored energy is available to humans.

Key points

- Food chain studies help in understanding feeding relationships and interactions and energy flow between organisms in any ecosystem.

- Organisms at each trophic level depend on those at the lower trophic level for their energy demands.

- Energy transfer between trophic levels follows the 10 per cent rule and the amount of energy available decreases at each successive trophic level.

- Food chain studies can help provide an understanding of the movement of toxic substances and biological magnification in ecosystems.

- Food chain studies can help in analysis of biological diversity in ecosystems.

Did you know?

Biomagnification of a pollutant occurs when the substance is:

- long-lived, because short-lived substances tend to be broken down before they can reach danger levels

- mobile since if it is not mobile it will remain in one place and may not be available to be taken up by organisms

- soluble in fats because such substances tend to be retained for long periods

- biologically active, a requirement for causing problems in organisms.

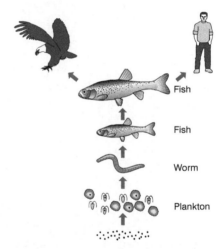

Location	DDT concentration/ppm	
	Decimal	Power of ten
humans	6.0	
fish (freshwater)	2.0	
fish (marine)	0.5	5×10^{-1}
aquatic invertebrates (freshwater and marine)	0.1	10^{-1}
freshwater	0.00001	10^{-5}
seawater	0.000001	10^{-6}

Figure 1.12.3 Biomagnification in a food chain

Learning outcomes

On completion of this section, you should be able to:

- explain what is meant by the 'productivity' of producers and ecosystems

- distinguish between different types of ecological pyramids

- explain energy and nutrient flows in ecosystems.

Productivity in ecosystem systems

Productivity is an important component of an ecosystem and refers to the rate of production of organic matter in any unit time in an ecosystem. Primary productivity is the rate of capture of solar energy or biomass production of producers in an ecosystem. Secondary productivity is the rate of assimilation or storage of food energy at the level of the consumers.

There are two types of primary productivity: gross primary productivity (GPP) and net primary productivity (NPP). GPP is the total rate of capture of solar energy or the total rate at which organic matter is produced by autotrophs. It is measured by chlorophyll/g dry weight/unit area or CO_2 fixed from dry weight/unit area. NPP is the rate at which autotrophs store organic material as new tissue. It is measured by g/m^2/day or kg/m^2/yr

$$\text{Net primary productivity} = \text{gross primary productivity} - \text{respiration}$$

$$\text{NPP} = \text{GPP} - \text{respiration}$$

Because only the energy that is stored in the tissues of autotrophs is available to consumers, NPP values are often used to compare the productivity of different ecosystems.

Ecological pyramids

The structure and function at successive trophic levels is often shown graphically by means of ecological pyramids. The first trophic level, the producer level, constitutes the base of the pyramid. Ecological pyramids are of three types: pyramid of numbers, pyramid of biomass and pyramid of energy.

A pyramid of numbers is a graphic representation of the number of individual organisms per unit area at each trophic level, starting with producers at the base of the pyramid. The shape of the pyramid varies from ecosystem to ecosystem.

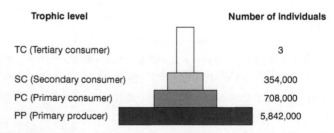

Trophic level	Number of individuals
TC (Tertiary consumer)	3
SC (Secondary consumer)	354,000
PC (Primary consumer)	708,000
PP (Primary producer)	5,842,000

Figure 1.13.1 *Pyramid of numbers in an ecosystem*

A pyramid of biomass is a graphic representation of biomass present per unit area of different trophic levels, with producers at the base and carnivores at the apex. Biomass is the total amount of living organic matter in an ecosystem at any time. In a terrestrial ecosystem, maximum biomass occurs in producers and progressively decreases from lower to higher trophic levels – thus the pyramid of biomass in terrestrial ecosystems is always upright.

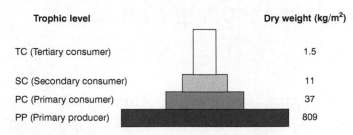

Figure 1.13.2 *Pyramid of biomass*

In aquatic ecosystems the pyramid of biomass is inverted because the biomass of the trophic level is highly dependent on the reproductive potential and longevity of each organism. An inverted pyramid is one where the combined weight of producers is smaller than the combined weight of consumers.

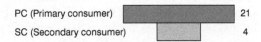

Figure 1.13.3 *Inverted pyramid of biomass in an aquatic ecosystem*

A pyramid of energy is a graphic representation of the amount of energy available per unit time and area in different trophic levels. Producers form the base and carnivores form the apex of this pyramid. A pyramid of energy is always upright because energy is transferred based on the '10 per cent' rule for the transfer of energy from one trophic level to another, as explained by Raymond Lindeman in 1942.

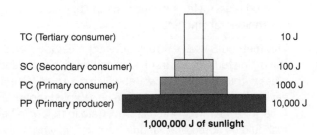

Figure 1.13.4 *A pyramid of energy*

Key points

- Pyramids of numbers and biomass may be upright or inverted depending on the nature of the food chains in a particular ecosystem.

- In most ecosystems producers are greater in number and biomass than the herbivores, and herbivores are greater in number and biomass than carnivores.

- Pyramids of energy are always upright.

- There is a gradual increase in energy at successive trophic levels from producers to consumers.

Population sampling – why do it and how?

There are times when we need to know what kinds of plants and animals are found in a particular habitat and how many of each species are present. Conducting population sampling helps to answer these two basic questions. One of the main assumptions made during population sampling is that the results obtained will be representative of the habitat. Population sampling is conducted using standard sampling units and methods to ensure that all of the samples are representative of the area, whether the habitat is terrestrial or aquatic. There are three main ways of taking samples:

Random sampling	Used when the area under study is uniform, large, there is limited time available and when there is a need to remove observer bias in the selection and taking of samples. Large numbers of samples are taken from different positions within the study habitat. Use is also sometimes made of a numbered grid and a random numbers table to determine where in the habitat the sample should be taken.
Systematic sampling	This method entails taking samples at fixed intervals. Transects, usually line transects or belt transects, are generally used in this approach. A sampling line is established across areas where clear environmental gradients exist. This method is useful for showing zonation of species along an environmental gradient.
Stratified sampling	This method is used to study different areas identified within the main study area. Stratified sampling involves identifying areas within the overall habitat which may be different from each other and therefore need to be sampled separately. This is because we rarely find uniform strata in the main habitat. It should be noted that random samples may not cover all areas of a habitat equally; stratified sampling can be used effectively in such conditions.

Ecological sampling techniques

It is generally impossible to sample entire populations in a given study area. It is therefore important to take smaller samples which will be representative of the area. One therefore has to consider how best to undertake the sampling taking into account a number of factors:

- what sampling method to use
- whether the organisms to be sampled are moving or non-moving
- the time available for the study.

Transects

Transect sampling is useful when investigating the transition from one community to another. The transect method is appropriate for use in areas that are zoned or have gradients. Two common types of transects are line transects and belt transects.

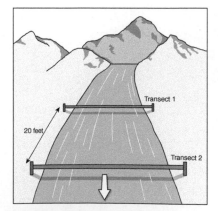

Figure 1.14.1 *How to construct a transect*

Line transect	Line transects are used when there is a need to investigate a possible linear pattern along which change in plant and animal communities is experienced. Compared to belt transects a line transect gives information on what is present but gives limited information on how much is present.
Belt transect	This method gives information on species abundance as well as presence or absence of species. This is similar to the line transect except that it is considered a widening of the line transect to form a continuous belt or series of quadrats.

Quadrat sampling

A quadrat is a frame of known size, typically 1 m^2 or 0.5 m^2. Quadrats can be used to estimate a population in an area which is fairly uniform. This is an appropriate method for estimating the abundance of organisms that are sedentary or non-mobile. They can produce three estimates of population size: density (the number of individuals per unit area in the habitat study site or the number of organisms per square metre), frequency (number of quadrats that contain the organism) and percentage cover.

Percentage frequency and percentage cover

Calculating percentage frequency requires a record of the number of samples and quadrats in which a species is recorded and the number in which it did not occur. Calculating percentage cover requires an estimate of the percentage area occupied by each species in each quadrat.

Two key advantages of using percentage frequency rather than percentage cover to estimate the abundance of a particular species are:

- The process is usually quicker.
- Percentage frequency requires the use of exact numbers while percentage cover relies on subjective estimates of the area.

Two main disadvantages of using percentage frequency rather than percentage cover to estimate the abundance of a particular species are:

- It does not take into account the size of individuals and may lead to incorrect conclusion about the importance of species in the ecosystem.
- It does not take into account abundance within each quadrat.

Key points

- To obtain an unbiased estimate of a population, the sampling procedure should be random.
- The use of transects is beneficial when investigating the transition from one community to another.
- The transect method is applicable for areas that are zoned or have gradients.
- Quadrats are appropriate for estimating the abundance of organisms that are sedentary or non-mobile.

Types of quadrats

There are two types of quadrats: frame quadrats and point quadrats. A frame quadrat is usually a metal, wooden or plastic frame of a known area that is used to isolate a subset of the population which comprises a single sample. Frame quadrats are useful for measuring population density, percentage cover or frequency of occurrence of a species. A point quadrat is useful for sampling stationary or non-mobile organisms. Point quadrats are used to record presence or absence of a species at a number of locations in the study habitat. Point quadrats are therefore useful for measuring percentage cover, and the proportion of ground covered and shaded by the aerial parts of organisms. Frame quadrats and point quadrats are also useful for estimating relative abundance using one of many abundance scales.

⌗ Links

Read 1.17 to get some more information on species abundance and abundance scales.

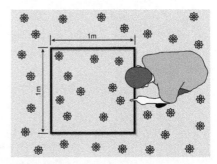

Figure 1.14.2 *How to use a quadrat*

Learning outcomes

On completion of this section, you should be able to:

- describe methods of sampling moving organisms
- discuss assumptions made when using the mark-release-recapture method to estimate population size
- identify abiotic parameters that should be measured when conducting ecosystem studies.

$$\frac{\text{Number marked in second sample}}{\text{Total caught in second sample}} = \frac{\text{Number marked in first sample}}{\text{Size of whole population (N)}}$$

Figure 1.15.1 *Equation used to estimate the population size (N)*

Sampling of moving organisms

Sampling animals may pose problems because either they are not all visible or they simply move about during the sampling exercise. There are methods available for sampling moving animal populations and one of these methods is the mark-release-recapture method.

Estimating population size using this method involves capturing the organisms, marking them in a non-harmful manner and then releasing them back into their natural habitat. The marked individuals are allowed enough time to integrate back into the population. A second sample is then taken and the numbers of marked and unmarked individuals are counted. Figure 1.15.1 shows the equation used to estimate the population size (N).

Assumptions made when using this method

A number of assumptions are made when using this method to estimate population size. These include:

- the proportion of marked individuals in the second sample will be the same as the proportion of marked individuals to unmarked individuals in the whole population
- organisms are captured randomly and there is no bias towards a particular group for both samples
- the mark is not lost between the period of release and recapture
- marking does not hinder the movement of organisms or harm them in any way or make them more or less likely to be preyed upon
- marked individuals mix randomly with unmarked individuals
- the likelihood of capture does not change with the age of the organism
- the organisms must not be trap-shy
- the population is a closed one in that during the study period, changes in population size due to immigration, emigration, death and birth are negligible.

Sample calculation: estimating the population of birds in a school garden

Here is a worked example calculating the population of the bird species *Pitangus sulphuratus* (kiskadee) in the garden of the Bushy Lot Secondary School.

During the a sampling exercise the following data were obtained:

- In the first sampling activity 10 birds were captured and marked with pit tags and released.
- In the second sampling activity 20 birds were caught. Four of these individuals were birds which were captured and marked in the first sample.

Using the equation in figure 1.15.1 the calculation is as follows:

$$\frac{4}{20} = \frac{10}{N}$$

so $N = \frac{20}{4} \times 10 = 50$

The estimated population size is 50

Estimating the size of deer population

In this procedure, a biologist uses traps to capture savannah deer alive and then mark them. The animals are returned unharmed to their habitat. Over a long period of time, the animals are continuously captured and a record is made of the marked individuals in the sample. Using the information presented, estimate the size of the deer population in the savannah.

Total number of deer caught	Number of deer recaptured	Number of marked individuals in second sample
120	80	30

Measuring abiotic factors in ecosystems

Abiotic environmental parameters influence distribution, abundance and activity of organisms. Local conditions such as temperature, rainfall and light intensity may affect terrestrial organisms. Water current, dissolved oxygen (DO), total suspended solids (TSS) and nutrient content may affect organisms in aquatic ecosystems. There are many field techniques available for measuring the range of abiotic parameters (physical and physico-chemical) in air, water and soil.

When conducting an ecological study some of these parameters are essential and important and should be determined. It is important to measure abiotic factors in the microhabitat of organisms when trying to relate the distribution and abundance of organisms to specific abiotic factors.

This table gives an indication of some of the abiotic environmental parameters that are determined when undertaking ecosystem studies.

Parameterà	pH	Temperature	DO	TSS	Conductivity	Light intensity	Turbidity
Terrestrial ecosystem	+	+			+	+	
Aquatic ecosystem	+	+	+	+	+	+	+
Fish	+	+	+	+			
Reptiles	+	+				+	
Birds		+				+	
Amphibians	+	+					
Plants	+	+	+				

Key points

- Studying animal populations is different from studying plant populations because animals are mobile and plants are non-mobile.
- The capture-mark-release-recapture method is useful for estimating population size.
- Environmental parameters influence the distribution, abundance and activity of organisms.
- It is important to measure abiotic factors in the microhabitat of organisms when trying to relate the distribution and abundance of organisms to specific abiotic factors.

On completion of this section, you should be able to:

- distinguish between primary and secondary ecological succession

- explain ecosystems as self-sustaining climax communities.

Ecological succession

Ecosystems are dynamic and biological communities are seldom static. Ecosystems change constantly in response to both abiotic and biotic factors. The relative abundance of species may change, and new species may enter the community while others may leave. There are several reasons for these changes, among which are:

- Catastrophes, which may have natural or anthropogenic origins.
- Seasonal changes which may be related to, among other things, changes in temperature, rainfall, light intensity and wind speed.
- Succession changes, which are long-term changes in composition of a community, brought about in large part by the actions of organisms in the community.

There are two types of ecological succession: primary succession and secondary succession. These succession changes take place in stages called seral stages which together, when complete, are called a sere. Each stage is characterised by its own community of organisms and its own abiotic conditions. During succession the environment is modified by each species, thus making it less suitable for the species present and more suitable for new species to colonise.

Primary succession

Primary succession takes place when the community develops on bare, uncolonised ground. This type of succession is typical of newly formed areas where no life existed previously, for example new land or rocks formed after volcanic eruptions, glacial deposits of rock, new ponds, and sand dunes. In these cases there is no pre-existing substrate and so soil formation is required. Species are then introduced into new areas that had never been colonised.

These first species are called primary colonisers or pioneer species. Pioneer species include lichens, algae and mosses, which are all adapted for living in harsh environments with low levels of nutrients and moisture. Over time organic matter builds up and more mineral particles accumulate, providing more suitable substrates for larger plants and a greater diversity of animal life.

Secondary succession

Secondary succession takes place in areas where there is soil and where life was already present but which experienced some alteration or environmental change. Disturbances can be either manmade or natural. In these conditions the changes are similar to those experienced in primary succession but differ only in that they occur at a faster rate. This is because of the initial presence of the soil and the seed banks that are present. One example of secondary succession is the series of changes observed after a forest fire decimates a forested area.

The process of succession

Once pioneer species establish themselves in an area, biotic factors become increasingly important. This is so because over time, plant and animal species change their environment, making it suitable for new species that replace their predecessors. Some of the important aspects of this process are listed below.

- Plant and animal species living in a particular location gradually change over time, as does the physical and chemical environment within that area.
- Ecological succession takes place because organisms interact with and affect the environment within a given area, gradually changing existing edaphic, biotic and biotic factors, through processes such as living, growing and reproducing.
- Each species is adapted to thrive and compete best against other species under a very specific set of environmental conditions. If and when these conditions change, species that are better adapted to the new conditions will outcompete and replace the existing species.
- A change in the plant species present in an area promotes a change in animal species present in the area. The reason for this is that each plant species will have different associated animal species that feed on it. The presence of particular herbivorous species will usually determine the specific carnivorous species that are present.
- Plants provide particular microhabitats which determine the structure of the plant communities as well as the animal species which can live in the given microhabitats.
- Fungal species associated with particular plant species will also change depending on the changes experienced with plant communities.
- The 'seres' are characterised by the presence of different plant and animal communities which gradually change from one sere to another. Although seres are not totally distinct from each other, one will tend to merge gradually into another, finally ending up with the final stage, the 'climax' community.

Climax communities are characterised by:

- An increase in species diversity and complexity of feeding interactions.
- A progressive increase in biomass.

Succession stages are named based on the conditions and environments under which they occur.

- Hydrosere describes succession in an aquatic environment.
- Xerosere describes succession that takes place on dry land.
- Halosere describes succession that occurs in a salty environment.
- Lithosere describes succession that takes place on rocky substrates.

Key points

- Biotic and abiotic factors in a habitat influence the species that live there.
- These factors impact and influence the gradual progression from one community to another – a process called ecological succession.
- Ecological succession is the gradual process by which ecosystems change and develop over time. Nothing remains the same for long and habitats are constantly changing.
- As succession progresses, the diversity and number of species increases and the feeding relationships become more complex.
- Eventually a stable climax community develops which is in equilibrium with the environment and which will undergo little further change.
- This climax community exhibits ecosystem stability, which is a measure of its sensitivity to disturbance or perturbation.

Did you know?

Edaphic factors relate to soil conditions and how they affect living organisms. Edaphic characteristics of soil include factors such as water content, acidity, aeration, and the availability of nutrients. Edaphic factors therefore refer to ecological influences on the properties of soil brought about by its physical and chemical characteristics.

Learning outcomes

On completion of this section, you should be able to:

- differentiate between species abundance, species diversity and species richness
- calculate species diversity.

A	Abundant
C	Common
F	Frequent
O	Occasional
R	Rare
N	None. Included when comparing sites and is used to indicate instances when none of the particular species is present.

Figure 1.17.1 *The ACFOR scale*

∞ *Links*

See 3.14 for more on the use of quadrats for sampling.

$$D = \frac{N(N-1)}{\sum n(n-1)}$$

Where:
D = Simpson diversity index
N = total numbers of individuals of all species present
n = total number of individuals of a particular species
$\sum$ = is 'the sum of'

Figure 1.17.2 *The Simpson species diversity index*

Species abundance, diversity and richness

Species abundance is the term used to express the total number of organisms found in a biological community. Species diversity is a measure of the number of different species, ecological niches, genetic variation or ecosystems present. Species richness is a measure of the number of species found in a sample location. There is an inverse relationship between the abundance of a particular species and the total diversity of a biological community. Communities with large number of species often have a small number of individuals of a given species in a particular area. Complexity in ecosystems refers to the number of trophic levels in the community and the number of species present at each trophic level.

Species abundance

Species abundance can be estimated using frame quadrats and point quadrats while making use of one of several abundance scales. Species abundance can be recorded by estimating the percentage cover or using the ACFOR scale where each species present at the study site is placed in one of the categories shown in Figure 1.17.2.

These scales are very arbitrary and highly subjective because different people may use different judgement criteria to place species in each category. ACFOR scales can be made less subjective if the criteria used for placing species in each category are standardised.

Species diversity

It is usually difficult to define species diversity because the definition consists of two distinct components: species richness – which is the variety of species or the number of different species in the community – and species evenness, which is the relative abundance of species. It is rare for all species to be equally abundant because some species are more competitive than others, while some have greater fecundity and thus are generally more abundant than others. Evenness therefore refers to how equally abundant each of the species are and it is a relatively simple way to combine abundance and richness.

Species richness

Species richness is an index that is based on the number of species and is a measure of the number of species found in a sample area. It is expected that the larger the sample, the more species will be present in the area. Species richness can be calculated from the number of species per unit area. It is simple to calculate but does not account for relative abundance and thus is not usually sensitive to environmental disturbance.

Species diversity is different from species richness in that it takes into account both the numbers of species and the dominance of species in relation to one another.

Calculation of species diversity

The species diversity index is the most commonly used measure of biodiversity. This is calculated using the species richness and the

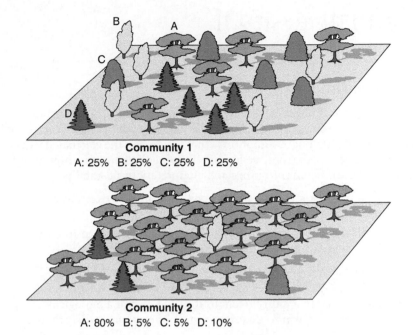

Community 1
A: 25%   B: 25%   C: 25%   D: 25%

Community 2
A: 80%   B: 5%   C: 5%   D: 10%

Figure 1.17.3 *Illustration of example of species diversity in two communities*

abundance of each species in the community. The Simpson species diversity index is one method used to calculate species diversity. The formula used to calculate the Simpson species diversity index is:

This index could be linked to the harshness of the environment or the level of pollution present in an ecosystem. In general, the species diversity is greater in environments where pollution levels are low and where abiotic conditions are not very harsh.

An example to calculate species diversity

In the example shown in the table each species is represented by a letter. Using the formula the species diversity can be calculated as follows:

$$D = \frac{25\,(24)}{(10 \times 9) + (8 \times 7) + (4 \times 3) + (3 \times 2)} = \frac{600}{164} = \mathbf{3.658}$$

In this example the species diversity calculated is 3.658

Species	Number of individuals
A	10
B	8
C	4
D	3
	Total individuals = 25

Key points

- Species diversity is a mathematical measure of diversity in an ecosystem.

- The more diverse an ecosystem the more complex it will be.

- Diversity indices provide information about rarity and commonness of species in a biological community.

- Quantification of species diversity helps us to understand ecosystem community structure.

- Species diversity differs from species richness in that it takes into consideration both the numbers of species present and dominance or evenness of species in relation to one another.

Learning outcomes

On completion of this section, you should be able to:

■ analyse the relationship between species diversity, community and ecosystem stability.

Diversity between species

Diversity of species in an ecosystem helps to protect it from threats. Natural selection causes populations to adapt to their environment and to differentiate from one another. Species differ from each other in their resource requirements and use, their environmental tolerance and their interactions with other species. Therefore species composition and interactions have a major influence on ecosystem functioning and stability.

Resilience and stability

Some biological communities tend to stay relatively stable and constant over time. The species diversity index is sometimes taken as an indicator of ecological stability. The assumption is that the greater the calculated valued for species diversity the more stable will be the ecosystem. It is important to understand the different meanings and interpretations of stability.

■ Ecosystem resistance refers to the ability of an ecosystem to resist change after a disturbance.

■ Ecosystem resilience refers to the ability of an ecosystem to return to its original state after experiencing some change.

■ Local ecological stability refers to the tendency of a biological community to return to its original state after a small or local disturbance.

■ Global ecological stability refers to the tendency of a biological community to return to its original state after a major disturbance.

While each of these types of stability is related to species diversity, this is not always the case. Thus it is now generally accepted that the relationship between species diversity and ecosystem stability probably varies with the environment. In fact one may conclude that healthy and simple biological communities in environments with very demanding abiotic conditions are likely to be more stable and less prone to human disturbance than biological communities that are fragile and complex but are found in environments with less demanding abiotic conditions.

Diversity within species

Species diversity is one component of the concept of biodiversity. Strictly speaking, species diversity is a measure of the diversity within an ecosystem. This is really the number of different species in a particular area, the species richness, weighted by some measure of abundance such as number of individuals or biomass. Another measure of species diversity is the relative abundance with which each species is represented in a given area – the species evenness.

Genetic diversity within species is very important when considering the conservation of biodiversity. Some reasons for this importance are:

■ It contributes to the species' ability to adapt to changing environmental conditions over time through natural selection and evolution.

■ It is important to the species' ability to colonise new areas and occupy new ecological niches.

■ Genetic diversity is thought to contribute to a species' ability to reproduce and produce robust offspring.

Species diversity and ecosystem stability

Why is ecosystem stability likely to increase as species diversity increases?

■ Ecosystems are interconnected by feeding relationships.

■ The more complex and interconnected a community is, the more stable and resilient it is expected to be in the event of a disturbance.

- If many different species occupy each trophic level some can fill in if others are stressed or eliminated from the community.
- This therefore makes the community resistant to change so that it can recover relatively easily from disruptions.

Why is ecosystem stability likely to decrease if species diversity decreases?

- In diverse and highly specialised ecosystems there are keystone species.
- Removal of these keystone species can result in the elimination of many other associated species.
- The feeding relationships would be disrupted as species may have died or species may have migrated.
- Loss of species diversity makes an ecosystem less resilient or less stable.

Significance of species diversity

Species composition in an ecosystem is the result of natural selection and evolution. Over time each species would have adapted to its own niche, characterised by certain specific factors including, for example, temperature range, light and availability of food, which enables the species to reproduce and thus maintain its population.

Being a component of the ecosystem in which it lives, each species interacts with its environment and performs certain specific functions. It is these interactions which are important for keeping the ecosystem in balance. Therefore any loss of species may affect the existence of other species and ultimately affect ecosystem balance.

When this happens many of the functions previously carried out will no longer be possible. However, although another species may take over the niche that becomes available it will be impossible for it to replace all of the functions performed by the previous species. This is important to remember because when a species goes extinct the services and functions that it provided to the ecosystem and biosphere are lost forever.

Anthropogenic impacts on species diversity

The loss of species is linked to a loss of ecosystem function, some aspects of which have serious implications not only for humans but also for the flora and fauna themselves. Loss of species can harm the quality of life of humans. It is therefore important for humans to do what is necessary to prevent loss of species.

Key points

- Complex ecosystems with high species diversity are assumed to be more stable due to the greater number of alternative feeding links between species.
- The higher the species diversity the greater the inertia and resilience of the ecosystem.
- In complex ecosystems several species may be able to carry out the same function. Loss of one species will therefore not result in serious compromise of ecosystem function.
- Two factors which affect ecosystem stability are biotic potential and environmental resistance.

Did you know?

Some actions of humans that can and do affect species diversity are:

- Overexploitation, pollution and habitat conversion are key threats to species diversity. These activities can impact species diversity at local, regional and global levels.
- Introduction of non-native species into ecosystems destroys delicate ecosystem balance.
- Humans contribute significantly to global warming, which has a substantial impact on ecosystems and species within ecosystems at the local, regional and global levels.
- Increasing use of natural resources, species and ecosystems to support various industries like tourism, agriculture and transportation also severely impacts species diversity.

Figure 1.18.1 *Human impact on an ecosystem: pollution of waterways*

Introduced species

Sometimes new species are introduced into areas where they normally do not occur. Such introductions may be either deliberate or accidental. Non-native species are sometimes very successful in their new environment. It is therefore important to pay attention to such introductions because interactions of introduced species with native species can have disastrous consequences for native species and the natural environment.

Reasons why non-native species are successful in new environments:

- Non-native species usually have high reproductive rates and short generation times. This helps their numbers to increase rapidly.
- In the case of plants, sometimes they introduce growth-inhibiting substances in the environment that prevent or restrict the growth of other species.
- Non-native species are usually generalists and so can survive better than other species under conditions in which other species will succumb.
- Non-native species usually have high genetic variability which makes them very adaptable to different environments and therefore they can survive under different environmental conditions.
- Usually they have high dispersal rates and use a variety of dispersal mechanisms to allow them to disperse over wide areas with varied environmental conditions.

Impacts of introduced non-native species:

- Promotion of migration of native species as the competition for resources increases.
- Some non-native species may be predators of native species and therefore an increase in predator–prey relationships in the environment may lead to the death and loss of native species.
- Introduced species often outcompete native species for resources.
- Increasing interactions and completion among and between species could disrupt feeding relationships and impact ecosystem stability.
- Non-native species could be important vectors of diseases.
- Non-native species could cause degradation of the environment due to different foraging and behaviour habits.

Intensity of species interaction

Interspecific competition is competition between members of different species for resources within an ecological community. Intraspecific competition is competition between members of the same species for resources within an ecological community. Intraspecific competition is more intense because all members of the species have identical resource requirements. This is because, unlike in interspecific competition, none of the competing organisms will seek alternative resources and therefore this will put severe pressure on the limited resources available in the environment. Competition for resources helps to keep numbers of a community within the carrying capacity of the environment and is therefore an important component of population control. It is also vital for the sustainability of natural resources needed for a particular species.

Species interaction and predator–prey cycles

Species interaction is important in ecosystems. Figure 1.19.1 illustrates an example of a typical predator–prey cycle. Why does this cycle show this particular trend?

Sample calculation of percentage frequency

An area of seashore in a Caribbean island was chosen and an ecological study was undertaken of the area by students preparing for their CAPE Internal Assessment. The study area was 60 metres from the upper to lower shore and 50 metres wide. The total area of the study site was 3000 m².

Eight quadrats were placed at intervals of 5 m along the length of a belt transect that was laid out at the study site. The size of the quadrats used was 0.5 m × 0.5 m. Results of the study are outlined in the table below.

Calculation of percentage frequency of plant species found along study site

Plant species	Quadrat number								Total	% Frequency
	1	2	3	4	5	6	7	8		
Species 1	+	+	-	+	+	+	+	-	6	75
Species 2	+	-	+	+	-	+	+	+	6	75
Species 3	-	+	+	-	-	-	-	-	2	25
Species 4	+	+	-	-	-	+	-	+	4	50

Calculating percentage frequency

Step 1: Count the number of quadrats in which the species is present

Step 2: Divide by the number of quadrats to get the frequency

Step 3: Multiply answer from step 2 by 100 to get the percentage frequency (% frequency)

Sample calculation for Species 1

Number of quadrats in which the species is present = 6

Frequency calculated = 6/8

Percentage frequency = 0.75 × 100 = 75%

Key points

- Interaction of introduced species with native species can have disastrous consequences for native species and the natural environment.

- Interspecific competition is competition between members of different species for resources within an ecological community.

- Intraspecific competition is competition between members of the same species for resources within an ecological community.

- Intraspecific competition is more intense because all members of the species have identical resource requirements.

- The numbers of the predator and prey are closely related. As predator numbers decrease the number of the prey tends to increase.

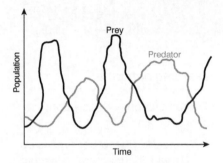

Figure 1.19.1 *A typical predator–prey cycle*

Did you know?

The cycle (Figure 1.19.1) shows this trend because of the following:

- If the prey population increases, there is more food for the predators so their numbers increase.

- More prey is then eaten, and so their numbers decrease.

- Predator numbers fall because there is not enough food.

- If the predator numbers are falling, the prey will become more numerous because they are not being eaten as much.

Calculate the percentage frequency for the remaining species and compare with the values in the table. You may wish to draw your own bar graph to display your results and compare with Figure 1.19.2.

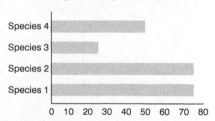

Figure 1.19.2 *Percentage frequency of plant species*

Learning outcomes

On completion of this section, you should be able to:

- state the observations made by Darwin about natural selection, evolution and adaptation
- explain the process of natural selection
- explain the role of natural selection in adaptation and evolution of species.

Adaptation, natural selection and evolution

What do we mean when we say that species are adapted to certain conditions? Adaptation is the ability of an organism to withstand extremes of physical conditions in the environment. Adaptation may be interpreted in two ways. The first is acclimation, a limited range of physiological modifications available to individual organisms. The changes that individuals acquire are neither permanent, nor are they passed on to their offspring. The second type of adaptation operates at the level of the population and is influenced by the inheritance of specific genetic traits that allow organisms to live in specific environments. Charles Darwin and Alfred Wallace developed the theory of evolution to explain this type of adaptation.

Darwin developed the theory of evolution by natural selection based on observations he made whilst sailing the world on the HMS *Beagle*.

Key observations made by Darwin were:

- All populations experience variation because organisms within populations exhibit individual variation in appearance and behaviour.
- Some traits that are heritable are consistently passed on from parents to their offspring. On the other hand there are some traits that are highly influenced by environmental conditions and show weak heritability.
- All populations have the ability to 'over-reproduce', leading to large population increases in the absence of limiting factors.
- The numbers of individuals in natural populations of organisms tend to remain relatively stable over long periods of time, despite over-reproduction of offspring.
- Individuals which are deemed the 'fittest' will survive and contribute more offspring to the next generation.

Darwin's theory of natural selection stated that:

- Individuals of the same species compete for the same resources since they have the same requirements and means of obtaining the said resources.
- Competition leads to high mortality and displacement rates due to predation, diseases and organisms dispersed into unfavourable environments.
- The individuals best adapted to obtain resources survive the competition, breed and pass on favourable genetic traits.
- This differential survival results in a population with individuals best suited to the existing environmental conditions for finding food, finding mates, escaping from predators and resisting disease. They therefore have a greater likelihood of being dispersed in favourable environments.

Role of natural selection in the evolution of species

- Natural selection acts on pre-existing genetic diversity.
- Selective pressures (due to limited resources or environmental conditions) favour advantageous genes.
- This leads to the survival of the fittest organisms which eventually reproduce successfully.
- Individuals with the favourable genes may have more offspring surviving into the next generation.
- These carry the favourable genes which are available for passing on to the next generation.

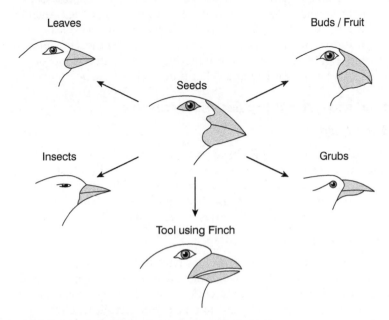

Figure 1.20.1 *Adaptive radiation shown by Darwin's finches*

Role of natural selection in adaptation of species

- Over the long term, populations of survivors gradually adapt to new conditions.
- For adaptation to occur there must be survivors.
- Selective pressures eliminate all individuals that cannot tolerate the new conditions.
- There must be enough survivors to maintain a viable breeding population.

Key points

- Natural selection and variation are important in the process of evolution.
- Natural selection and adaptation can cause organisms with similar origins to be very different in appearance and develop different habits over time.
- Natural selection affects individuals, while evolution and adaptation operate at the level of populations.
- Natural selection requires heritable variation and differential survival and reproduction associated with the possession of a heritable genetic trait.

On completion of this section, you should be able to:

- outline the benefits of natural ecosystems

- describe anthropogenic impacts on ecosystems and biodiversity

- justify the need to maintain ecosystem integrity.

Humans and natural ecosystems

Humans depend on some components of natural ecosystems for survival. This is essentially because humans are connected to both the living and nonliving parts of the ecosystem. The human economy is dependent on the services provided by natural ecosystems.

Natural ecosystems perform and support essential life-support services for human civilisation as we know it. Some of the key services include air and water purification, functioning of biogeochemical cycles, decomposition of wastes, regulation of climate, restoration of soil fertility, maintenance of biodiversity and pollination – which all support our agricultural, pharmaceutical and other industrial activities.

Natural ecosystems provide resources for:

Food and medicines

Indigenous and other peoples have used products from natural ecosystems for centuries to provide food and medicines. Forest ecosystems have provided wildlife for food and soil for the cultivation of agriculture crops. Coral reefs and seagrass beds have supported commercial and subsistence fisheries for generations.

Recreation and aesthetic benefits

Natural ecosystems and their resources have been used to support tourism activities in the Caribbean for years. Wildlife, coral reefs and beaches have all been used for recreational activities in the Caribbean. Natural ecosystems have religious and aesthetic significance for indigenous and other communities.

Figure 1.21.1
Recreational benefits:
a Caribbean beach

Revenue generation and industries

Natural ecosystems provide the raw materials for many industries. Through the use of the raw materials in industries many Caribbean countries generate revenue for the people and the country's economy. Some examples include timber and non-timber forest products, fisheries and minerals from the mining industry.

Protection to infrastructure and human life

Natural ecosystems (coral reefs, beaches, seagrass beds and mangroves) are important coastal ecosystems in the Caribbean that offer protection to coast communities and associated infrastructure. Without these ecosystems, Caribbean islands and low-lying coastal communities would be subjected to the constant onslaught of waves. Their presence helps to reduce monetary spending on engineering reinforcements.

Contribution to global and local climatic conditions

Caribbean countries have forest resources, and the trees absorb carbon dioxide and release oxygen. Intact forest ecosystems help to reduce the enhanced greenhouse effect when carbon dioxide is absorbed by the trees. Natural aquatic ecosystems in the Caribbean also help to absorb carbon dioxide and so contribute to the reduction of the enhanced global greenhouse effect.

Genetic resources

Natural ecosystems provide genetic resources that are useful as medicines and for enhancing agricultural crop production. The use of natural genetic resources for medicines and for enhancing food productivity and security helps to save lives and reduce the dependence on artificially produced medicinal and food products.

Figure 1.21.2 *Coral reefs*

Why should we maintain ecological integrity?

It is important to maintain ecological integrity since this helps to:

- maintain species, ecosystem, and habitat diversity and stability
- promote continuous and efficient cycling of matter and flow of energy
- limit competition for scarce resources
- maintain and sustain ecosystem productivity.

Ways in which humans disrupt the integrity of natural ecosystems

- improper disposal of non-biodegradable products
- point and non-point pollution from industries and other activities
- removal of litter and recyclable organic materials from the ecosystem.

Key points

- Natural ecosystems perform and support essential life-support services for humans.
- Natural ecosystems provide resources for food and medicines, recreation and aesthetic benefits, revenue generation, infrastructure protection, genetic resources and global and local climatic conditions.
- It is important to maintain ecological integrity in order to continue making use of the benefits provided by natural ecosystems.
- Humans can disrupt the integrity of natural ecosystems by introducing different forms of stressors into ecosystems.
- Humans compromise the stability and functioning of ecosystems through pollution, damage to habitats and disruption of species diversity and abundance.

On completion of this section, you should be able to:

- identify factors affecting population growth in a natural ecosystem

- distinguish between density-dependent factors and density-independent factors affecting population growth

- explain how biotic potential and environmental resistance affect population growth.

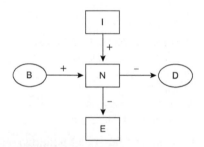

Figure 1.22.1 *Factors that alter population density*

Terms that describe population change:

- **population size:** the number of individuals in a population

- **population density:** the number of individuals per unit area

- **population growth:** a change in the number of individuals in a population. Population growth is positive when the overall number of individuals in a population increase and negative when the overall number of individuals in a population decrease

- **population growth rate:** the change in number of individuals in a population per unit time. This could be positive or negative.

Populations and population change

Recall that a population is a group of organisms of the same species living together in one geographical area at the same time. Within a population, each individual capable of reproduction has the opportunity to reproduce with other mature members of the group. There are four ways in which changes are brought about in the numbers of individuals in a population. These four ways are:

- births (B)
- immigration (I)
- deaths (D)
- emigration (E)

Populations gain individuals when young are born or new members join the group from other populations. Immigration is common in mobile animal species. In plants, immigrants are usually associated with spores and seed dispersals.

Population growth under ideal environmental conditions

Figure 1.22.2 shows the typical growth curve for a population colonising a new area where initial conditions are ideal: there is abundant food, no competitors, no predators or diseases; there is optimal temperature, among other favourable environmental factors. The growth curve may be described in stages or phases (see table), each stage characterised by specific responses to the prevailing conditions.

Phase 1 or Lag phase	Growth curve rises slowly (slow growth rate) because there are few reproducing individuals and it is difficult to find mates because of low population density.
Phase 2 or Log phase	Population grows at its biotic potential (maximum rate). Birth rate exceeds death rate. Environmental factors begin to impact population growth amidst increased competition for resources (environmental resistance).
Phase 3 or diminishing growth phase	Intensity of the effect of environmental factors on population growth increases and the population growth rate slows down. The growth curve becomes less steep.
Phase 4 or stationary phase	Population growth rate steadies and tapers off. Carrying capacity is reached. Resource availability stabilises the population size.

Factors that affect population growth

In natural ecosystems environmental factors interact to either increase or decrease population size.

Factors that increase population size	Factors that decrease population size
Adequate food resources	Limited food resources
Adequate water resources	Limited water resources
Adequate habitat space	Inadequate habitat space
Ability to withstand diseases	Inability to withstand diseases
Very little predation or good ability to escape predators	High level of predation or inability to successfully avoid predators
High reproductive rates	Low reproductive rates
Generally stable abiotic conditions	Generally unstable abiotic conditions

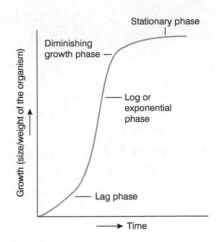

Figure 1.22.2 *Phases in population growth*

Density-dependent and density-independent factors

The factors that affect population growth can be classified as either density-dependent factors or density-independent factors. Density factors are usually biotic and include food availability, level of predation and ability to resist disease.

Density-independent factors are usually abiotic factors and affect the same proportion of the population irrespective of the population size or density. Density-independent factors include weather events and natural disasters that affect populations.

Biotic potential and population growth

Populations differ in their capacity to grow. The biotic potential of a population is the maximum rate at which it can increase when resources are unlimited and when environmental conditions are ideal. Each species has a different biotic potential due to differences in fecundity, including:

- age of reproductive maturity
- 'litter size'; the number of offspring produced at each reproductive event
- number of reproductive events that occur in an individual's lifetime
- 'survival rate of the species'; how many offspring survive to reach reproductive age.

Environmental resistance and population growth

Populations cannot continue to grow exponentially indefinitely. Environmental limits exist, called environmental resistance factors, which impact the number of individuals capable of surviving and reproducing in a given habitat at any one time. These factors are either density-dependent or density-independent.

Why should we care about changing population numbers?

Populations that are increasing or decreasing can provide indicators of potential problems occurring in the organism's environment. Ecologists always need to know why population numbers in an area fluctuate.

Key points

- Increases in population numbers are due to births and immigration while decreases in population numbers are due to death and emigration.
- Exponential growth occurs when only biotic or intrinsic factors affect a population.

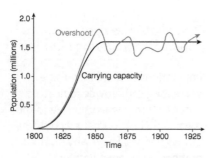

Figure 1.23.1 *Population increase and carrying capacity*

Carrying capacity and human populations

Is there a carrying capacity for humans and, if so, what is it and when will it be overshot? The concept of carrying capacity can be applied to humans but it should be noted that because the consumption habits and patterns of humans are variable it is difficult to predict the earth's carrying capacity for humans.

Carrying capacity

Environmental carrying capacity is the maximum population size that can be sustained by a particular environment over a relatively long period of time. If there is overpopulation there can be many implications for the environment. Carrying capacity may be considered to be the theoretical equilibrium population size at which a given population will stabilise in a particular environment when its resources remain constant. It is the maximum sustainable population size that can be supported indefinitely without degrading the habitat and limiting resources for future members of the population and for future generations.

There are instances when population increases are rapid during the exponential growth phase, resulting in an overshooting of the carrying capacity (Figure 1.23.1). Under such conditions the environment will be unable to support the growing population. Populations that show such growth curves are called boom-and-bust populations. In such instances, overpopulation can harm the environment and lead to a lower carrying capacity.

Factors that affect carrying capacity

Based on the concept of carrying capacity, all populations have limits to growth in nature. Expanding populations always reach a size limit imposed by certain factors such as water, space and nutrients or by prevailing adverse conditions such as extreme weather events (drought, floods and temperature extremes). Environmental resistance together with biotic potential keeps a population in balance and prevents it from overshooting the environmental carrying capacity.

Two schools of thought on carrying capacity

Ester Boserup: It is generally acknowledged that carrying capacity for humans is a function of population size as well as differing levels of consumption, and types of technologies used in production and

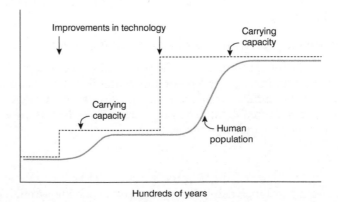

Figure 1.23.2 *Concept of the relationship between technology and carrying capacity according to Ester Boserup*

consumption. It should be noted that technologies that impact overall consumption also impact carrying capacity (Figure 1.23.2). This school of thought is based on the work of Ester Boserup, a Danish agricultural economist who suggested that, as populations increased, we would, out of necessity develop technological solutions to increase our resource base, for example developing artificial fertilisers to grow more crops. Therefore we do not reach carrying capacity because the resource base has been developed.

Further, when consumption levels per capita are high then a smaller population can be supported. The question that one may ask is, can we keep developing new technologies and abilities to exploit and use new resources that promote increases in carrying capacity? Another important question is, at what price will such developments be realised, relative to our quality of life?

Thomas Robert Malthus: Thomas Robert Malthus was known for his theories about change in populations. According to his theory, human population exhibits a J-shaped curve and is accelerating (Figure 1.23.3). Malthus argued that population increases exponentially, but resources increase linearly and so eventually we exceed carrying capacity.

An examination of this diagram and an understanding of the concept of carrying capacity according to Malthus would tell us that such growth cannot continue because resource limitations and/or environmental degradation will cause human population growth curves to approach an upper limit, the carrying capacity.

The carrying capacity for humans is acknowledged to be more complex and the definition is usually expanded to include not degrading our cultural and social environments and also not harming the physical environment in ways that will adversely affect future generations. This implies and links with the concept of sustainability and sustainable development.

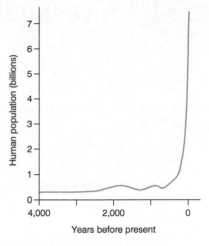

Figure 1.23.3 J-Shaped population growth based on Malthus's theory

'Carrying capacity is determined jointly by human choices and natural constraints. Consequently, the question, how many people can the Earth support, does not have a single numerical answer, now or ever. Human choices about the Earth's human carrying capacity are constrained by facts of nature which we understand poorly. So any estimates of human carrying capacity are only conditional on future human choices and natural events.' Joel Cohen

Cohen, Joel. How Many People Can the Earth Support? New York: W. W. Norton, 1995.

Key points

- For a given area, the carrying capacity is the maximum number of individuals of a given species that could be supported by the resources present without significantly degrading or depleting the resources.

- The carrying capacity may be lowered by resource destruction and degradation during overshoot events or extended through technological and social changes.

- It is important to note that most resources that humans use are finite and we are constrained as are all other organisms.

- It is logical that human carrying capacity is dependent on our lifestyles and quality of life.

- Human carrying capacity must also be thought of in cultural terms (cultural carrying capacity).

On completion of this section, you should be able to:

- investigate different parameters in terrestrial ecosystems
- discuss some environmental parameters in terrestrial ecosystems
- extract, present and interpret data collected in ecological studies.

Terrestrial and aquatic ecosystems

Terrestrial and aquatic ecosystems have some common elements and also some that are different. These parameters often characterise the specific ecosystem and it is therefore important for these parameters to be measured. Environmental parameters in the microhabitat of organisms often shed light on the distribution and abundance of organisms.

Similarities between terrestrial and aquatic ecosystems	Differences between terrestrial and aquatic ecosystems
In both ecosystems there are communities made up of a variety of species.	Aquatic ecosystems are at times more stable, with smaller fluctuations in temperature and other variables.
Both ecosystem types have populations at different trophic levels.	Organisms in aquatic ecosystems are rarely exposed to desiccation.
There is a great amount of interdependence between species in both ecosystems.	Oxygen is more often a limiting factor in aquatic ecosystems. This is seldom the case in terrestrial ecosystems.
Stratification occurs in both ecosystems.	Light can be a limiting factor in aquatic ecosystems.
Ecological succession takes place in both types of ecosystems.	Organisms in terrestrial ecosystems are influenced more by gravity. Organisms in aquatic ecosystems are usually supported by the water.

Parameter	Role of factor
Light energy from sunlight	This is the primary source of energy in almost all ecosystems. Visible light is important for photosynthesis. Quality of light, light intensity and length of the light period are important aspects of light as an abiotic factor.
Temperature	Extremes of temperature influence the presence or absence of species and their functioning. Temperature influences opening and closing of flowers, germination of seeds, breaking of dormancy and migration patterns of animal species.
Water	Water is essential for life and all organisms depend on it for survival. Terrestrial organisms can suffer from desiccation in limited water and plants show adaptation to water availability by being adapted as hydrophytes, xerophytes or mesophytes.
Total suspended solids (TSS)	TSS includes all suspended particles in water which do not pass through a filter. As TSS levels increase water loses its ability to support a variety of aquatic organisms. Suspended solids absorb heat from sunlight, eventually increasing temperature and reducing dissolved oxygen levels in water. Suspended solids also physically impact aquatic organisms by smothering and suffocating them.
Dissolved oxygen (DO)	The amount of oxygen dissolved in a body of water and which is available to sustain aquatic life. DO levels are influenced by water temperature, rate of photosynthesis, degree of light penetration, water turbulence and the amount of oxygen used in respiration and decay of organic matter.

Atmospheric gases	Oxygen is important for respiration. Nitrogen is important for plant growth. Carbon dioxide is used for photosynthesis by plants.
Winds or air current	Winds play a role in pollination and dispersal in plants as well as some animal species. Plants exposed to strong winds are usually smaller than those in less windy conditions. Wind is an important erosion factor that can remove or redistribute topsoil.
Soil (edaphic factors)	Soil texture, soil temperature, soil water content and soil pH, soil organisms and soil decaying matter are all important in plant and animal populations' presence or absence.
Physiographic factors	The physical nature of the area, these factors include altitude and slope. Altitude plays a role in vegetation zones. Slopes affect soil surface temperature.

Presenting data from an ecological study of a terrestrial ecosystem

A group of students conducted an ecological study of a forest plot in a Caribbean country. The forest plot has many species of trees that herbivores feed on. The table below gives a summary of the number of different animal species found in general taxonomic groups identified during the study. The table also shows examples of each of the groups that were identified.

Group of organism	Number identified	Examples	Percentage of total number of groups identified
Mammals	15	tapir, agouti, jaguar	19
Birds	20	kiskadee, hawk, harpy eagle	26
Fishes	10	piranhas, houri	13
Insects	25	butterflies, moths, grasshopper	32
Reptiles	8	lizard, snake	10

1. Draw a bar chart to show the data presented in the table to show the different groups of organisms identified in the forest plot.
2. Draw a pie chart to show the percentage of the different groups of organisms that were identified in the forest plot.
3. Use the information in the table to construct a food chain for the forest plot.

Different examples can be drawn from the information in the table. One example is:

Kiskadee → hawk → harpy eagle

Note that the head of the arrow points in the direction of the organism that does the eating.

Key points

- The way in which plants and animals grow and carry out their functions is the result of several abiotic factors.
- These factors include light, temperature, pH, water, wind, soil and physiographic factors.
- Data collected in ecological studies can be presented in different ways.

Learning outcomes

On completion of this section, you should be able to:

- investigate different parameters in aquatic ecosystems
- discuss some environmental parameters in aquatic ecosystems
- extract, present and interpret data collected in ecological studies
- discuss elements of developing a monitoring plan.

Investigating a coastal wetland ecosystem

A group of students conducted an ecological study to compare different mangrove ecosystems in a Caribbean country for their CAPE Internal Assessment. During this study both moving and non-moving organisms were sampled and the results were used to make conclusions about the diversity and stability of the two locations. Students used the transect method to conduct their study of plant communities. Results are given in the table below.

Species	Number of plants at Site A	Number of plants at Site B
Avicennia germinans	10	25
Laguncularia racemosa	30	15
Rhizophora mangle	5	45
Conocarpus erectus	40	20

Objective 1: Candidates had as one of their objectives the task of displaying the results of their study. A bar graph was used to display the results. Try drawing a bar graph of the results.

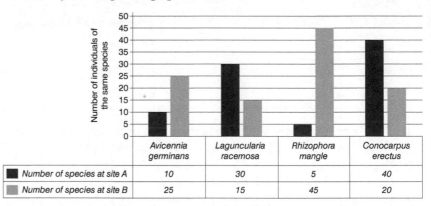

Figure 1.25.1 Numbers of each plant species at study sites

Objective 2: Candidates had as a second objective the task of comparing the species diversity of the two sites and commenting on the species abundance and ecological stability of the two sites. The species diversity for Site A was calculated to be 2.81. The first step in making this determination would be to calculate the species diversity for Site B.

Solution: Calculation of species diversity for Site B

$$D = \frac{N(N-1)}{\sum n(n-1)}$$

$$= \frac{105\,(104)}{25\,(24) + 15(14) + 45\,(44) + 20\,(19)}$$

$$= \frac{10,920}{600 + 210 + 1980 + 380}$$

Did you know?

When drawing a graph marks are usually awarded for:

- the title of the graph
- correct labelling of each axis
- choice of an appropriate scale for each axis
- plotting of the points correctly
- drawing a proper curve or bar for all points.

$$= \frac{10,920}{3,170} = 3.44$$

The species diversity of Site A (2.81) is less than then the species diversity of Site B (3.44).

Some points to note when considering species diversity and ecological stability:

- Ecosystem stability refers to the ability of biological communities to remain relatively stable and constant over time.
- Ecosystem stability is linked to the variety of feeding relationships in the ecosystem.
- Ecosystem stability tends to increase as species diversity increases.
- Since ecosystem stability increases as species diversity increases then the ecosystem with the highest species diversity will most likely be the most stable.
- Ecosystems are interconnected by feeding relationships and therefore the more complex and interconnected a community is, the more resilient it is expected to be in the event of a disturbance.
- In diverse and highly specialised ecosystems keystone species exist and their removal can result in the elimination of other associated species.
- Disruption of feeding relationships makes an ecosystem less resilient or less stable.

Objective 3: Students had to choose an appropriate method for undertaking the study and outline how the method was implemented as their third objective.

An example of an appropriate method and how the method is applied: Line transect:

- A transect line can be made using a rope marked and numbered at specific intervals (0.5 m or 1 m intervals), all the way along its length.
- This transect is laid across the area to be studied.
- The position of the transect line is very important and it depends on the direction of the environmental gradient you wish to study.
- A line transect is carried out by unrolling the transect line along the gradient identified.
- The species touching the line may be recorded along the whole length of the line (continuous sampling).
- Alternatively, the presence, or absence of species at each marked point is recorded (systematic sampling).
- If the slope along the transect line is measured as well, the results can then be inserted onto the profile.

Key points

- Ecosystem stability refers to the ability of biological communities to remain relatively stable and constant over time and is linked to the variety of feeding relationships in the ecosystem.
- The more complex and interconnected a community is, the more resilient it is expected to be in the event of a disturbance.
- In diverse and highly specialised ecosystems keystone species exist and their removal can result in the elimination of other associated species.

Some points to note when considering species diversity and species abundance:

- Species diversity is a measure of the number of different species in an ecosystem.
- Species abundance is an expression of the total number of organisms in a biological community.
- Species abundance is determined by the relative abundance of individuals in the biological community.
- The abundance of a particular species is inversely related to the total diversity of the community.
- Relative abundance refers to the evenness of distribution of individuals among species in a biological community.
- Communities with a very large number of species have only a few members of any given species in a particular area.

Learning outcomes

On completion of this section, you should be able to:

- investigate different parameters of an aquatic ecosystem
- extract, present and interpret data collected in ecological studies.

Investigating aquatic pond ecosystems

Ponds are very good aquatic systems for ecological investigation. In this study students will measure selected physical and chemical parameters of two ponds and collect bottom-dwelling (benthic) invertebrates from each pond. Students will use the information collected to compare the patterns of community structure for each pond given the observed abiotic and biotic conditions of the two ponds.

This activity is intended to familiarise students with the aquatic ecosystems of ponds and the diversity of life found in them. Students will also measure select physical and chemical parameters of the pond and examine their role in influencing the distribution and abundance of organisms in the pond.

Objective 1: To measure abiotic parameters and then construct a table with the information obtained.

Surface area	Estimate the surface areas of the ponds. For circular ponds measure or estimate the diameter. For rectangular ponds estimate or measure the length and width. Determine the depth of the pond by using a calibrated line with weight attached.
Temperature	Measure and record the temperature of the air and the temperature of the surface water with a thermometer. Measure all temperatures in the shade.
Transparency	For shallow ponds take measurements where the bottom is easily viewed from the surface. In deeper ponds, use a Secchi disc to measure and determine transparency.
pH	Measure the pH after standardising the pH meter. Always follow instructions for the use of the pH meter.
Nitrate, phosphate and sulphate	Follow directions for individual test kits to measure the concentrations of these ions in water samples.
Total dissolved solids	Follow directions for individual test kits to measure this parameter.
Dissolved oxygen	Follow directions for individual test kits to measure this parameter.

Objective 2: To investigate the bottom-dwelling (benthic) biotic diversity of sample ponds.

Organism	POND 1 Number of individuals	POND 2 Number of individuals
Caddisfly larvae	15	9
Crustaceans	8	25
Dragonfly nymphs	5	1
Damselfly nymphs	9	5
Hydra	3	0
Nymphs of true bugs	70	60
Water striders	10	0
Total of organisms collected	**120**	**100**

Objective 3: To calculate the relative abundance of each type of organism collected from the pond using the equation:

$$\frac{\text{number of individuals of a species}}{\text{total number collected of each species}} \times 100 = \text{relative abundance}$$

Sample calculation for Pond 1: water striders: (10 individuals/120 total) × 100 = 8.3%

The table is completed with remaining calculations.

Organism	POND 1 Relative abundance (%)	POND 2 Relative abundance (%)
Caddisfly larvae	12.5	9
Crustaceans	6.7	25
Dragonfly nymphs	4.2	1
Damselfly nymphs	7.5	5
Hydra	2.5	0
Nymphs of true bugs	58.3	60
Water striders	8.3	0
Total	**100**	**100**

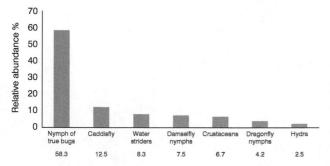

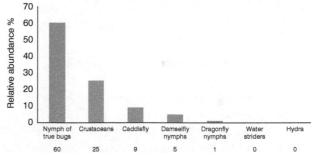

Figure 1.26.1 *Graph of relative abundance vs. type of organism for Pond 1*

Figure 1.26.2 *Graph of relative abundance vs. type of organism for Pond 2*

Key points

- Ponds are very good aquatic systems for ecological investigation.
- The information collected can be used to compare the patterns of community structure given the observed abiotic and biotic conditions.

Multiple-choice questions

1 Intraspecific competition:

   A is less intense than other forms of competition.

   B occurs within a community but not within an ecosystem.

   C occurs between individuals of the same species.

   D occurs between individuals of different species.

2 What percentage of energy is lost when it is passed from one trophic level up to the next?

   A 90 per cent

   B 60 per cent

   C 20 per cent

   D 10 per cent

3 If a keystone species disappears from an ecosystem:

   A populations of the other species may decrease dramatically.

   B only their common prey will be affected.

   C only their typical predator will be affected.

   D Both A and B are correct.

4 The phosphorus cycle differs from the carbon cycle in that:

   A there is little or no human impact on the phosphorus cycle.

   B phosphorus is not a critical component of living organisms.

   C the atmosphere does not contribute to part of the phosphorus cycle.

   D plants play a role in the carbon cycle, but have no role in the phosphorus cycle.

Questions 5 and 6 refer to the following table, which shows the total number of different fish species in an aquatic ecosystem in a Caribbean country.

Common name of fish	Total number of organisms
Tilapia	22
Houri	28
Sunfish	40
Hassar	3
Silverfish	7
Total number of organisms:	**100**

5 Calculate the diversity index (D) using the following formula:

$$D = \frac{N(N-1)}{\sum n(n-1)}$$

   A 2.50

   B 3.28

   C 3.50

   D 4.50

6 What percentage of the pond community is made up of tilapia and houri?

   A 25 per cent

   B 28 per cent

   C 50 per cent

   D 35 per cent

7 Which of the following organisms would be first to colonise an area during primary ecological succession?

   A grass

   B lichens

   C herbs

   D shrubs

Questions 8 and 9 refer to the following diagram:

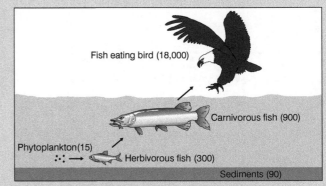

Figure 1.27.1 *Aquatic ecosysem showing mercury concentration (ppm)*

8 The process shown in the diagram is:

   A bioremediation

   B biomagnification

   C bioaccumulation

   D phytoremediation

9 The concentration factor responsible for the different amounts of the mercury pollutant at the various levels in the food chain is:

A 15

B 325

C 1200

D 12000

10 Environmental resistance may limit the size of populations by:

A increasing both birth and death rates.

B decreasing both birth and death rates.

C increasing death rates and/or decreasing birth rates.

D decreasing death rates and/or increasing birth rates.

Essay questions

1 a Define:

    i ecosystem

    ii ecosystem stability   [4 marks]

  b The following organisms were identified in a terrestrial ecosystem:

Grass	Hawk	Kiskadee	Caterpillar	Grasshopper

    i Use ALL of the organisms given in the list to construct a food web for the terrestrial ecosystem.   [5 marks]

    ii Describe two ways in which humans can disrupt the integrity of the terrestrial ecosystem.   [8 marks]

    iii Explain the importance of decomposers in the terrestrial ecosystem.   [3 marks]

                   Total 20 marks

2 The following table presents the results of a study by a student who investigated how populations of a predator and prey changed over time. The data was gathered over a six-month period.

Species numbers at study site in sugar cane estate

Species	Month 1	Month 2	Month 3
Predator	35	20	10
Prey	25	15	15

Species	Month 4	Month 5	Month 6
Predator	10	15	25
Prey	20	25	15

a Draw an appropriate graph to present the data shown in the table.   [8 marks]

b With reference to your graph describe the trends displayed by BOTH predator and prey over the period. You must include specific values in your answer.   [4 marks]

c Describe the role of intraspecific competition in resource partitioning in ecosystems.   [4 marks]

d Using appropriate examples, distinguish between 'gaseous biogeochemical cycles' and 'sedimentary biogeochemical cycles'.   [4 marks]

                   Total 20 marks

3 Snail kites feed on freshwater snails. Figure 1.27.2 shows the variation in population numbers of both snails and snail kites for an eight-month period, January to August, of one year.

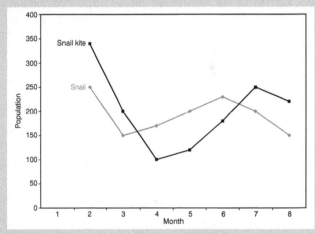

Figure 1.27.2 *Population variation for snails and snail kites*

a With reference to Figure 1.27.2, describe the trend in variation for the population size of BOTH organisms for January to August.   [4 marks]

b Explain how predation by snail kites could benefit the snail population shown in Figure 1.27.1.   [4 marks]

c Explain how a named density-independent factor may impact the population size of snails.   [4 marks]

d Outline TWO possible impacts on the ecosystem if the population of snails is significantly reduced by draining of the waterway.   [8 marks]

                   Total 20 marks

2.1 The relationship between people and the environment

People and the environment

Human beings are found over much of the Earth's surface in a variety of environments and have generally adapted their way of living to deal with the physical conditions they encounter. Technology is used to enable people to survive in conditions which may otherwise prevent existence, as in the case of extreme cold or heat.

The poor are often more dependent directly on the environment, as in the use of rain as irrigation, or the planting of seeds saved from the previous year's harvest. As such they are more vulnerable to environmental changes, so a failure of the monsoon means a failure of crops, or a poor harvest one year means fewer seeds for the next year. In more affluent countries, the environment is often a backdrop to human activity rather than an integral part of their existence – as in the highly urbanised environment where temperature and humidity are controlled electronically rather than relying on natural breezes and air flow. Another example is the use of machines and technology which mimic the outdoors so that one can exercise by running or cycling through scenic locations without leaving the comfort of the gym!

In extreme environmental conditions, human beings will adapt their behaviours in order to survive and maintain their livelihoods. Ways in which they do this include:

- wearing protective clothing to protect themselves, for example coats in cold conditions and loose, flowing robes in hot areas

- minimising outdoor activities at the hottest time of the day

- growing crops that are best suited to the local climatic conditions, for example growing rice rather than wheat in wet climates

- constructing shelters to protect themselves from heat or cold, rain or wind

- engaging in recreational activities which take advantage of the local environmental conditions, for example surfing in beach-oriented cultures, hiking and camping in national forests.

People have not only adapted to their environments through their behaviours, but even, over time, by changing their body types. Thus, those who live at high altitudes with thinner air, such as the Indians in Peru, have a greater lung capacity than those nearer sea level, which allows them to function in their local environment. People living in Arctic environments tend to be smaller and shorter, to avoid heat loss from their bodies as much as possible, compared with taller, leaner people in tropical environments.

People also have the ability to adapt to changing environmental conditions. A response to an immediate change in the environment is known as acclimatisation.

For example, people travelling from New York to Colorado must become acclimatised to the higher altitudes. Their bodies undergo some readjustment in order to cope with the thinner air, and in this

Acclimatisation refers to a short-term, reversible physiological response to a change in environmental conditions.

process they often suffer from altitude sickness – experiencing nausea, headaches and blurred vision. Where change is more permanent, adaptation takes place. This is achieved over a significantly longer time span, from generation to generation, as in the physiological changes from Neanderthal to *Homo sapiens sapiens*.

It is also significant to note that while humans have for the most part adapted to their environment, with increasing levels of technology they have also had the opportunity to change their environments to better suit their requirements. Thus, grasslands or savannahs have been maintained by burning to provide for the cattle which are kept there, while rainforest trees have been harvested for their timber and other forest products, or cleared for agriculture.

As humans became more settled and grew in number, their impact on the environment became more noticeable than the impact of the environment on them. So, slash and burn agriculture, where farmers move through the rainforest clearing plots as natural soil fertility declines, has been replaced by the large-scale clearing of forest to produce farmland, and the use of inorganic fertilisers to increase crop production.

Dependence on ecological systems and processes

Despite the impact of human activities on the environment, ecological systems are the basis of existence for most organisms, including humans, who rely on these processes for food, water, recreation and cultural activity. From the air we breathe to the crops we grow and eat, we depend on these systems and processes. Ecosystems act to:

- Regulate the climate – through the regulation of global temperature, as well as precipitation at the local and global level.
- Provide a consistent water supply to ensure the storage and movement of fresh water through watersheds, various reservoirs and aquifers.
- Cycle nutrients – all nutrients have to be recycled so they are made available for uptake by plants and animals.
- Regulate populations – in the form of actions of predators to maintain or control populations of species and avoid overabundance of animals or plants.
- Provide genetic resources – made available to humans for medicines, materials for manufacturing, ornamental varieties.
- Provide areas of recreation – lakes, rivers, forests.

The value of the services provided by ecosystems is largely unrecognised until there is a disruption or disturbance that has an economic cost for humans. For example, a coral reef's contribution to human livelihood through the provision of fish or through its ongoing provision of sand for beaches or protection from storm waves is not appreciated until the reef is destroyed or degraded by land- or sea-based pollution, leading to the loss of beaches and fish as well as coastal erosion and flooding, which have an impact on our daily lives. Similarly, the role of forests in helping to regulate the water cycle is belatedly recognised after flooding occurs way downstream.

Major threats to ecosystems come from human activities such as land use changes which reduce biodiversity or interfere with nutrient cycling, the introduction of exotic species, and pollution. Human activities already have a widespread impact on ecosystems at all scales and in all locations, since our activities have global as well as local implications. The most irreversible impact is the loss of native biodiversity.

Adaptation refers to a long-term, generational change.

Did you know?

Many ecosystem services are not valued in currency, and there is no warning in the form of an increase in price, for example, when these services are threatened or degraded, unless the result of that degradation changes its availability for use by humans. Thus, there is a lack of awareness of the impact of human activities until it is well advanced and difficult to reverse. Thus, rainforests continue to be cleared, until soil erosion, flooding and loss of biodiversity are impossible to reverse.

Key points

- Humans have changed their behaviours to live under varying climatic conditions.
- The physical environment has also been altered by human activities.
- Humans depend heavily on ecosystem services for their very existence.
- Ecosystem services are largely undervalued by people until they become degraded or are lost.
- The value of ecosystem services to human populations is immeasurable.

On completion of this section, you should be able to:

- define the terms 'abiotic' and 'biotic'
- differentiate between abiotic and biotic factors
- describe the impact of these factors on population distribution and activities.

Persons per square mile
0   1   100   500
0   0.4   39   193
Persons per square mile

Figure 2.2.1 *Population distribution in Guyana*

■ Dense tropical forest
■ Seasonally flooded tropical forest with scattered marsh and swamp
■ Grassland or savannah
□ Cultivated land

Figure 2.2.2 *Vegetation types in Guyana*

Abiotic factors

Abiotic factors are nonliving factors which affect living organisms in terrestrial and aquatic systems. They can be categorised as meteorological (temperature, precipitation, winds, sunlight, humidity); soil (pH, nutrient availability); water (salinity, pH). These factors vary over time and space to have an impact on the organisms. As seen in Module 1, section 1.4, abiotic factors tend to be density-independent and result in population changes due to factors such as storms, floods, relief and extreme temperatures. Where people settle reflects the choice of environments that present the greatest opportunity for growth. Alluvial lowlands, for instance, show one of the highest rates of population density due to the favourable physical conditions for the production of food – flat land, fertile soil, and natural irrigation from the flooding of rivers. The coastal lowlands of Guyana, for example, support the majority of that country's population (90%) while the Nile delta in Egypt is home to over 40 million people, half of that country's population. In contrast, marginal lands with poor soils are used by a significantly smaller number of people because they are unstable and infertile, and cannot produce enough food to support a large population.

Comparing maps showing areas of population density or distribution with various abiotic factors such as rainfall, vegetation or soils will illustrate clearly the relationship which may exist between these factors.

The rainforest covering much of Guyana, has influenced the settlement of the population. Thick, heavy vegetation results in low population density. In areas where the land is more readily available for use as farmland, e.g. in the coastal lowland areas, population densities are higher.

Swamps and marshlands are also avoided as the land is not useful for cultivation unless it is drained. There would also be problems with mosquitos and other insects (disease), reducing the attractiveness of the area for settlement.

The location of resources which may be used as a source of livelihood is also a factor; hence, there is a town at Linden (which floods seasonally) due to the presence of bauxite.

Abiotic factors play an important part in determining the location of settlements, and ultimately how humans make use of their environment. Climate is one significant factor which determines human location and activities. Where climates tend to extremes (too hot/cold/wet/windy), settlement is limited.

It is evident from Figure 2.2.3 that population densities are highest along the coasts, most clearly seen in North America and China, while areas subject to extreme weather and climates are very sparsely populated. Examples of such regions are Northern Canada, Russia and Australia. The dependence of humans on abiotic factors such as water availability is very clearly seen in the population distribution along the Nile River in Egypt. The high population concentration in the Ganges delta in India and Bangladesh also reflects the importance of fertile soils, and the presence of nutrients, to human settlement.

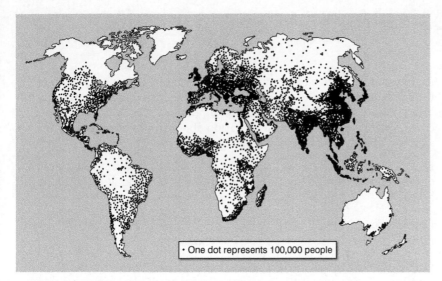

Figure 2.2.3 *World population density*

Biotic factors

Biotic factors are living factors which affect living organisms in terrestrial and aquatic ecosystems. They include factors such as predation, disease and parasites. They also play a role in population regulation and determining where humans engage in particular activities. The presence of the tsetse fly, which causes trypanosomiasis in humans and cattle, is a biotic factor which hinders settlement in certain areas of Africa.

CASE STUDY

Tsetse fly impact on human activities

In Nigeria, while the number of rural household cattle increases proportionally with the amount of land brought into the cultivation cycle, the positive association between cropping and cattle becomes weaker as exposure to trypanosomiasis increases. In areas with a trypanosomiasis prevalence of more than 30 per cent, it becomes virtually impossible to establish and maintain a mixed-farming system.

The significant reoccupation of fertile valleys has triggered a major expansion of agricultural production. It is also evident that the presence of tsetse prohibits effective and productive use of natural resources, however, by constraining the integration of livestock with crop production. The alleviation of this constraint is a matter of some urgency because of the extremely high and increasing demand for land resulting from the demographic explosion, a situation that is particularly acute in this subregion. These high population pressures also influence the movement of traditional pastoral societies away from the Sudano-Sahelian zone.

Source FAO 'The tsetse fly and its effects on agriculture in sub-Saharan Africa'

The presence of mosquitos which spread disease in the human population is a major biotic factor which affects where people settle. Swamps and marshlands are largely avoided due to the presence of these insects, coupled with the physical abiotic conditions which may exist there. For instance, the area of the Florida Everglades is very sparsely settled, or the Great Morass in Jamaica.

Key points

- Abiotic and biotic factors influence population distribution.

- Abiotic refers to nonliving factors such as climate and soils.

- Biotic refers to living factors such as competition and predators.

Populations are studied by examining their basic components; that is, their makeup by age and sex. The age and sex structure of a country can help to forecast trends with respect to population growth or decline of the country, and it can also give information about important relationships such as dependency ratios within a country.

Dependency ratio refers to the percentage of the population too young or too old to work who are dependent on those of working age, as seen below:

$$\text{Dependency ratio} = \frac{\text{\# of persons aged below 15} + \text{\# of persons aged over 65}}{\text{\# of persons aged 15–65 years}} \times 100$$

People of working age are also referred to as the economically active members of the population. A high dependency ratio suggests that the economically active population is low; this has serious implications for economic growth as well as the provision of services in the country. A smaller working population means reduced production, unless productivity is very high. It also means a reduction in taxes which are collected by the government, leading to restrictions on the services which can be provided for the very young (schools, nurseries) and the very old (health care, home assistance programmes). In Europe, the greying population means that the younger workforce must develop high levels of skills in order to keep production at the same level (or higher) once the existing workers grow old and retire. The use of technology can mitigate some of the loss of workers, but it means that young people must be very highly educated and skilled.

The age and sex structure can also be used to determine the population's median age, which divides the population in half, with as many people above the median age as there are below it. A high median age suggests that the population is aging, while a low median age signals a young population, which will likely grow quite rapidly in the years to come.

For instance, a median age of 18 means that the country has a huge young population, which will move upward into the reproductive ages. This has serious implications not only for population growth, but also employment, housing and the availability of services.

Country/region	Median age in 2010
Caribbean	*29.6*
Barbados	37.5
Belize	21.8
Guyana	23.8
Haiti	21.5
Jamaica	27.0
Trinidad	30.8

Source United Nations

Country/region	Median age in 2010
Europe	*40.1*
Germany	44.3
Albania	30.0
Asia	*29.2*
Japan	44.7
Afghanistan	16.6
Africa	*19.7*
Mauritius	32.4
Uganda	15.7

The composition of a population by gender can be an indication of the cultural attitudes towards women in society. A skewed gender ratio often favours a higher percentage of males as compared to females. In both China and India, where the cultural preference is for boys, the gender ratio is skewed in favour of males. This has severe implications, especially for China, where a generation of men will be hard-pressed to find wives as 30 million more men than women are expected to reach adulthood by 2020.

Age groups

In demographic studies, age is usually collated into age groups or cohorts of five years. This grouping facilitates the visual representation of the population in a population pyramid.

Doubling time

The rate at which populations grow has implications for a number of decisions, from government planning and provision of social services, to the rate at which resources may be consumed. With reference to those issues, the doubling time is a critical measure. It is defined as the amount of time in which a given population will double, based on its annual growth rate.

A rapid doubling time means that the population will double in size very quickly, which puts a huge strain on existing resources, including the provision of food, infrastructure and services.

Growth rates around the world

Africa	Rate	Asia	Rate	Europe	Rate	Caribbean	Rate
Liberia	4.5%	Singapore	3.52%	Luxemburg	2.086%	Turks and Caicos	4.56%
Mauritius	0.66%	Japan	0.023%	Slovakia	0.171%	US Virgin Islands	0.056%
Average	2.3%		1.08%		0.203%		0.718%

Source United Nations

Example: With a growth rate of 4.5%, Liberia's population will double in 15.6 years, while Mauritius's will take 106 years to double.

The faster the rate of doubling, the less time that governments have to put measures in place to deal with increased populations. Where the population is very young, the requirements will be mainly for child health provision, as well as schools and nurseries. Ultimately, there will be a high demand for jobs as these young people reach working age.

Some countries with longer doubling times would have different issues. The slow population growth would have repercussions for economic growth and activity, including the increase in the dependency ratio as workers reach retirement and are not replaced, as well as a declining domestic market due to a decrease in population size.

Population growth rates will often determine the natalist position taken by governments – to encourage or discourage growth.

Did you know?

The typical gender ratio is 105 males to 100 females. In China it is 122 to 100, and in some regions may reach 130 to 100.

∞ Links

More information on pro- and anti-natalist policies is given in 2.10 and 2.13.

Key points

■ Age and sex data can help predict future population change.

■ The relationship of one age group to another determines the dependency ratio.

■ Some countries have a gender imbalance due to population control policies.

■ Doubling time is dependent on annual population growth rate.

2.4 Demographics: Fertility, mortality and lifespan

Learning outcomes

On completion of this section, you should be able to

- define the terms 'fertility rate', 'mortality rate', 'lifespan' and 'life expectancy'

- understand the difference between lifespan and life expectancy

- understand the impact of fertility rate on overall population growth

- distinguish between mortality rates at different ages.

⟲ Links

See 2.6 for an explanation of the Demographic Transition Model.

Fertility rate

Fertility rate refers to the average number of children that a woman will have throughout her reproductive life. It is a useful measure of the rate of growth of a population. The higher the average fertility rate, the more children each woman is likely to have, and the faster the population will grow. The number of live births per 1000 population is the birth rate, and is related to the fertility rate. However, the fertility rate is usually given with reference to a particular age group, while the birth rate tends to be a crude measure for the entire population.

A drop in the fertility rate is one of the signs that the country is moving through the Demographic Transition Model, and the birth rate is falling. As countries become more developed, women tend to delay marriage and childbearing in favour of further education and careers, and the fertility rates decline. Later marriage means that women start having children later in their lifetime, and this reduces the potential number of children they will have. A woman who starts having children at the age of 15 is likely to have more children than one who starts at the age of 35.

One of the most important measures of fertility for the population growth of a country is the replacement fertility rate. Given that only women can bear children, they must have on average two children who will replace their parents. This rate is seldom equal to 2.0 exactly, as within any population there are women who will have no children, as well as some who will have more than the replacement rate, and the value of the replacement fertility rate reflects these differences. In the Caribbean as a whole, the replacement fertility rate is approximately 2.1.

In countries with very high fertility rates, population growth is rapid; family sizes are large, and on the whole, if this occurs in developing countries, the health of the children and their mothers will suffer. Where children are born in rapid succession, maternal health usually suffers and each successive child is then placed at a distinct disadvantage.

Mortality rate

Mortality rate refers to the death rate; the number of people dying per 1000 in the population at the current time. Death rates are often given as crude rates, which do not take into account the age distribution of the population. Therefore, it is not unusual for a developed country like Japan to have a crude death rate of 8.8, while a developing country like Guyana has a crude death rate of 5.9. However, Japan has an aging population so it is not unusual for it to have a high death rate. Guyana's high death rate may be the result of poor quality of life, or inadequate access to health care.

The death rate can be further subdivided into the infant mortality rate, which is the number of babies who die between birth and the age of one year, as well as the child mortality rate (the number of children between the ages of 0 and 5 who die).

Countries with a high degree of conflict will tend to have higher mortality rates in particular age groups, such as those from which soldiers tend to be drawn. An age and sex pyramid for Germany would show a lower than expected number of males aged 60 and above due to the deaths of soldiers in the Second World War. In places where health care is limited or sanitation is inadequate there may be higher rates of infant mortality as small children are among the most vulnerable in the population.

Death rates are therefore a function of both health and sanitation, as well as the age structure of the population.

Lifespan and life expectancy

Life expectancy is an estimate of the *average* number of additional years a person could expect to live if the age-specific death rates for a given year prevailed for the rest of his or her life, while lifespan refers to the number of years an individual is expected to live. Life expectancy therefore can change according to age group. A cohort of persons may have one life expectancy at birth, given the conditions prevailing at that time but, having survived to an older age, their life expectancy may change again as they reach further milestones.

Globally, the average life expectancy in developed countries ranges between 80 (New Zealand) and 83 (Japan), while in developing countries, it may be as low as 44 years (Central African Republic). Life expectancy reflects conditions with respect to health care and sanitation in a country. It gives some indication of the way in which the people of that country live or are cared for. Where life expectancies are low, generally there are issues related to war, environmental disasters such as drought and famine, as well as poor sanitation or inadequate supplies of clean drinking water.

Key points

- For each indicator listed, it is possible to separate them according to age. These age-specific results give a better overall picture of conditions in the country.

- All three measures also relate in some way to social conditions in the country.

- Fertility rates can have a significant impact on other measures in the country.

- Developed countries do not have low mortality rates, except in younger age groups; this may be contrary to what is expected. However, high death rates are a feature of aging populations, as commonly found in developed countries.

 Links

See 2.6 for more on the age and sex pyramid.

 Exam tip

When interpreting graphs with multiple lines representing different years or commodities or factors, highlight each line with a different colour highlighter/crayon so that at a glance, you can trace each line's path clearly.

On completion of this section, you should be able to:

- understand the term 'immigration'
- understand the term 'emigration'.

Immigration

The movement of people from one area to another to establish a new settlement is known as migration. People migrate to different places within the same country (internal migration), or they may move from country to country (international migration) for a variety of reasons. The movement of people into an area is referred to as immigration, and it serves to increase the population of a region or country. Most migration is voluntary, and people move to find a better life.

Many countries with low population growth rates leading to an aging population have used immigration as a means of ensuring a labour supply which will maintain economic growth, but there are often social and cultural conflicts when immigrants' beliefs and ways of life clash with those of their host country. This was seen in France where the wearing of the full face veil common in Muslim culture was banned in public places, resulting in severe conflicts between Muslim immigrants and local French laws.

Most immigration flows from developing countries to developed countries, as people leave their homes in search of jobs, or for a better way of life. Some immigration may be seasonal, as workers, especially young men, leave home for a few months each year to earn money, such as farm workers in Canada or miners in South Africa. This can create a number of social problems in the sending country as families are broken apart and wives are left to shoulder the burden of maintaining family farms in addition to supporting the family until their husbands can send money home. In the Caribbean many children were left behind to be raised by relatives as 'barrel children' in Jamaica when their parents migrated for a better life to the USA or UK.

Immigration deals with the short-term labour shortage due to a decline in population growth, but may raise issues about the way in which migrants become a part of their new societies. They may be expected to either assimilate or integrate, and the ways in which they do so may come into conflict with members of their host communities. Assimilation usually means that the immigrant wholly adopts the culture and customs of their new country, while integration results in the acceptance of the differences brought by the migrants to the society. Conflict usually arises where migrants are expected to assimilate, but would rather integrate or even set themselves apart from the norm.

This can therefore mean that immigration results in problems between immigrants and resident populations, and many societies with declining populations may encourage pro-natalist policies, rather than immigration, as a way to increase population growth.

Emigration

Emigration refers to the movement of people out of one country and into another. It deals therefore with the impact on the sending country, in the same way that immigration refers to the problems of the receiving country.

Given that most migration occurs when people leave poor countries for more economically developed ones, the impact is made that much harder

∞ Links

More information on pro- and anti-natalist policies is given in 2.10 and 2.13.

LEDC stands for less economically developed country. MEDC stands for more economically developed country.

on the sending countries. Most migrants are young and better educated than their peers, and therefore are the ones which a poor country is ill-equipped to lose. This brain-drain deprives these developing countries of the skills, knowledge and ideas needed to help the country become more developed.

Migrants are usually the ones who are leaving in search of a better life and may tend to be young men, or young families. They may move to large cities in search of better jobs and services, leaving their sending countries bereft of members of the economically active population, thus increasing the dependency ratio.

One of the benefits of emigration is the slowing down of the sending country's population growth rate. Since the migrants tend to be those in the reproductive age groups, it means that the sending country has the opportunity to reduce its population growth quite significantly.

Mass emigration from the Caribbean to the UK and USA in the 1950s and 1960s caused an overall reduction in the population growth in the region. The Caribbean's population growth rate therefore is quite low overall, with an average between 1 and 2 per cent.

Did you know?

The Caribbean has lost more than five million people to emigration. The countries with the greatest losses include Guyana, Suriname, Jamaica and St Lucia.

☑ Exam tip

Read the question you are answering very carefully and look for the key terms such as STATE or EXPLAIN. These words guide you as to the way in which you must answer, as well as the depth of your answer. 'State' means that you simply give a definition, or an example. If you are asked to 'explain', you should account for why a situation is the way it is.

Example

- State TWO reasons for emigration from LEDCs to MEDCs.

Response: 1) The search for jobs 2) Civil war or conflict

- Explain ONE effect of emigration on LEDCs

Response: Emigration from LEDCs means that the dependency ratio is increased, since it is usually the younger, economically active members of society who leave. It is more difficult for the remaining workers to support those who are not working, i.e. the very young and the elderly.

Key points

- People move in and out of countries in search of better lives.
- Immigration can help maintain economic activity where population growth is low.
- Immigration can result in social issues, causing problems.
- Emigration relieves population pressure.
- Emigration can lead to 'brain-drain' in LEDCS.
- Migration can have major impacts on families.

Learning outcomes

On completion of this section, you should be able to:

- recognise a population pyramid
- construct a population pyramid
- analyse a population pyramid
- understand the fertility rate
- understand the mortality rate
- understand the birth rate.

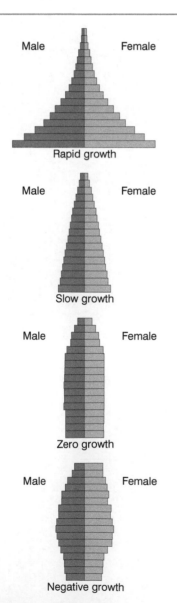

Figure 2.6.2 *Types of population pyramids*

Age and sex structure: The population pyramid

Demographic information is most often presented in an age and sex pyramid, also known as a population pyramid. The data are shown in two bars, side by side, with males on the left and females on the right. The data may be in form of percentage of the population or absolute numbers.

Reading a population pyramid

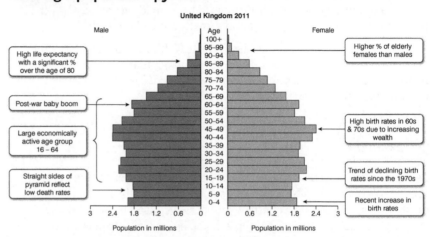

Figure 2.6.1 *Example of a population pyramid for the United Kingdom 2011*

How to read a population pyramid:

- The width of the base gives an indication of the birth rate of the country; a wide base indicates a high birth rate while a narrow base means that the birth rate is low.

- Asymmetrical pyramids indicate that there is a difference in the male and female populations. Depending on the stage at which this asymmetry is found, this could be due to loss in wars, or life expectancy.

- The shape of the sides also conveys information about death rates. The more concave the sides, the higher the death rate, while straighter sides indicate a low death rate.

- Irregularities or bumps in the sides may indicate anomalies such as baby boom periods, or an influx of migrant workers.

Population pyramids are related to the stages of the Demographic Transition Model (DTM) as illustrated in Figure 2.6.3. The changing shapes represent the changes in population structure as countries develop, moving from the high birth and death rates of the least developed countries, where population growth is slow, through the population explosion phase where death rates have fallen while birth rates remain high, ultimately to a hypothetical stage in which death rates exceed birth rates, leading to a declining population.

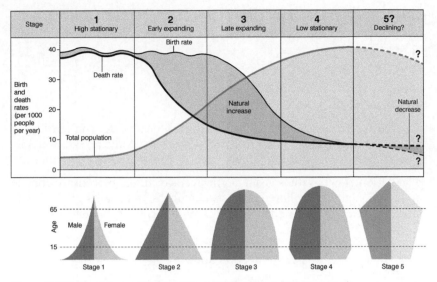

Figure 2.6.3 *The demographic transition model and population pyramids*

Other demographic data

Fertility rate, mortality rate and birth rates are other forms of demographic data which can give an overview of population status and predict population trends into the future.

Fertility rate is often given as the Total Fertility Rate, which refers to the average number of children a woman would have through her reproductive life. Two children per woman is the accepted replacement fertility rate; a rate higher than this means that the population is increasing, while one lower than two suggests a population in decline. Higher rates tend to be seen in developing countries, especially among the least developed countries where rates of 5.9 (Burundi) or even 7 (Niger) are not unheard of. In the developed countries, especially in Europe, rates are often well below two, leading to declining populations, for example Germany (1.4) and Sweden (1.7).

Mortality rates are given as the Crude Death Rate, which gives the overall number of deaths per 1000 of the population, regardless of age group. This indicator is affected by age distribution, and tends to show an increase in developed countries, as the population ages – for example Germany's mortality rate is 11. It is also high in countries where health care is inadequate, as the infant and child mortality rates will also tend to be quite high. Countries at war or engaging in civil conflicts will also show high mortality rates, such as Sudan (17).

Birth rates are given as Crude Birth Rates per 1000 of the population. This data can be further broken down by age group. The birth rate is the major factor which determines overall population growth rate along with the fertility rate and age structure. For example: Niger (46.8), France (12.6), Singapore (7.9).

Key points

- Population pyramids show the age and sex structure of a population.

- Changes in population can be easily seen on a population pyramid.

- Fertility, mortality and birth rates significantly impact population growth.

Learning outcomes

On completion of this section, you should be able to:

- calculate population growth rates
- calculate migration rates
- calculate birth and death rates
- calculate percentage population change
- calculate doubling time.

Population growth rate

Population size is a function of the birth rate, death rate, and fertility rate. Currently China (1,349,585,838 people) and India (1,220,800,359 people) are the two most populous countries in the world.

Population growth rate is given as the average annual percentage change in the population: the difference between births and deaths plus/minus the balance of migrants entering and leaving a country. The formula is:

$$\frac{(\text{Births} - \text{deaths}) +/- \text{net migration}}{\text{total population}} \times 100$$

Worked example 1:

$$\frac{(3500 - 2000) + 400}{250,000} \times 100 = 0.76$$

It can also be given as:

$$\frac{\text{birth rate} - \text{death rate}}{10}$$ but this represents **natural increase** only.

Worked example 2:

Birth rate: 16/1000; death rate 8/1000.

Growth rate = 16 – 8/10 = 0.8

Given the population growth rate, the population size at a given time can be calculated.

Worked example:

The population of Barbados in 2008 was 272,200. The rate of growth of population was 0.2%. The population in 2009 was therefore: 272,744.

Population growth: 0.2/100 × 272,200 = 544

Final population = 272,200 + 544 = 272,744

Percentage increase in population

The final population after all changes are taken into account is given by the following formula:

P1 + (B – D) + (I – E) = P2,

where

P1 = Early population   D = Deaths

P2 = Later population   I = Immigration

B = Births   E = Emigration

Where deaths and emigration are higher than births and immigration, then the later population will reflect a decline in population, while the reverse means population growth.

To calculate percentage population change, find the difference between P2 and P1, then find the percentage change as a function of the starting population as shown below.

Fertility rate

There is the general fertility rate, which measures the average number of live births per 1000 women aged 15–49 in any given year.

$$\text{Formula: } \frac{\text{number of births}}{\text{number of women aged 15–49}} \times 1000$$

Fertility rates can also be calculated for specific age groups to see differences in fertility at different ages or for comparison over time. This can help governments to target family planning policies.

$$\text{Formula: } \frac{\text{number of births to women in a particular age group}}{\text{number of women in that particular age group}} \times 1000$$

The total fertility rate (TFR) is the average number of children that would be born to a woman by the time she ended childbearing if she were to pass through all her childbearing years conforming to the age-specific fertility rates of a given year.

The replacement rate is the average number of children a couple must have to replace themselves in the population. It is standard as 2.0, but there are instances where the replacement rate would be higher or lower than 2, due to the fact that most people in the population are either *not* having children at all, or are having many children.

Migration rate

This can be given by immigration and emigration as a rate of population change.

Immigration rate is the number of immigrants arriving per 1000 population at a destination in a given year, while emigration is the number of migrants leaving a destination per 1000 population.

Net migration is the difference between immigration and emigration, while the net migration rate is the net migration given as a rate per 1000 population.

$$\text{Formula: } \frac{\text{number of immigrants – number of emigrants}}{\text{total population}} \times 1000$$

Migrants can also have an impact on population growth rates, especially if they are young and in their reproductive years.

Key points

- Birth, death and fertility rates can be calculated for entire population or by age group.
- Migration can have a significant impact on overall population change.
- Doubling time can have a major impact on population growth.

Birth rate

The crude birth rate is the number of babies born per 1000 of the population. There are also age-specific birth rates, which measure how many babies are actually born by age group.

The formula to calculate birth rate is:

$$\frac{\text{Number of births}}{\text{Total population}} \times 1000$$

Did you know?

Crude death rate is the number of deaths per 1000 of the population per year.

$$\text{Formula: } \frac{\text{Number of deaths}}{\text{Total population}} \times 1000$$

Learning outcomes

On completion of this section, you should be able to:

- understand the changes in population taking place over time

- account for the changes in population growth over time.

Historical trends

Trends in human population

Historically, growth in human populations has been very slow. However, over the past 30–40 years, it has taken less time to add each successive billion as shown in the table.

World population milestones		
World population reached:	Year	Time to add 1 billion
1 billion	1804	
2 billion	1927	123 years
3 billion	1960	33 years
4 billion	1974	14 years
5 billion	1987	13 years
6 billion	1999	12 years

Source: United Nations Secretariat, Department of Economic and Social Affairs, The World At Six Billion (1999)

The world population reached 7 billion in 2011, and it had taken just 12 more years to add another billion.

Large, often catastrophic changes in populations are seen on this population timeline, which illustrates the rise and fall of human populations. A significant fall in population occurred during the progression of the Plague, or Black Death as it was called, which swept Europe in the 1300s.

Through the early decades of the Industrial Revolution, life expectancies were low in western Europe and the United States. Thousands of people died from infectious diseases such as typhoid and cholera, which spread rapidly in the crowded, filthy conditions that were common in early factory towns and major cities, or were weakened by poor nutrition. But from about 1850 through 1950, a cascade of health and safety advances radically improved living conditions in industrialized nations. Major milestones included:

- improving urban sanitation and waste removal;

- improving the quality of the water supply and expanding access to it;

- forming public health boards to detect illnesses and quarantine the sick;

- researching causes and means of transmission of infectious diseases;

- developing vaccines and antibiotics;

- adopting workplace safety laws and limits on child labor; and

- promoting nutrition through steps such as fortifying milk, breads, and cereals with vitamins.

By the mid-20th century, most industrialized nations had passed through the demographic transition. As health technologies were transferred to developing nations, many of these countries entered the mortality transition and their population swelled. The world's population growth rate peaked in the late 1960s at just over 2 percent per year (2.5 percent in developing countries).

Demographers currently project that Earth's population will reach just over nine billion by 2050, with virtually all growth occurring in developing countries (Fig. 8). Future fertility trends will strongly affect the course of population growth. This estimate assumes that fertility will decline from 2.6 children per woman in 2005 to slightly over 2 children per woman in 2050. If the rate falls more sharply, to 1.5 children per woman, world population would be 7.7 billion in 2050, whereas a slower decline to 2.5 children per woman would increase world population to 10.6 billion by 2050.

Annenberg Learner (www.learner.org) The Habitable Planet, Unit 5: Human Population Dynamics, Section 4: World Population Growth Through History

Did you know?

When the Black Death struck Europe in 1348, the population of England fell by 20% in three years. Life expectancy of 32–35 years was reduced to under 18 years in medieval England.

Current and future trends

Even though most women are having fewer babies than their grandmothers, due to previously high levels of fertility, many women are still entering their reproductive ages. This means that birth rates will remain high, despite the overall fall in fertility, and population growth is not expected to decline until at least 2025. Therefore, the rate at which fertility declines will have a major impact on future population growth.

Europe's population is already declining. Population growth is expected to be the highest in sub-Saharan Africa, where fertility rates have fallen more slowly than anywhere else.

Most populations will also be aging, which is the result of falling fertility and increased life expectancy. This is a problem for developed as well as developing countries. In the Caribbean, with the population following the global trend in aging, elderly populations are expected to increase to 18 per cent of the total population in 2050. In 2000, six countries in the region had more than 10 per cent of their population over 60 years old. By 2025, both Barbados and Cuba are projected to have 25 per cent of their populations over 60.

✓ Exam tip

You should allow yourself approximately 20–25 minutes to complete a question. An essay outline is recommended to help organise your points and ensure a smooth, flowing response, rather than a disorganised and jumbled one.

Did you know?

Only one in 10 people lived in cities in 1900. By 1994 the figure had grown to one in every two people, creating megalopolises with millions to tens of millions of inhabitants. More than 400 cities have a population of over one million people.

Look here for more: didyouknow.org/population

Key points

- Human population growth was largely negligible for the first few thousand years of history.
- Billions have been added to the global population in increasingly shorter increments of time.
- A global total population of 9 billion people is projected by 2050.
- Overall growth is largely dependent on fertility rates, especially in developing countries.

Population distribution in developing nations

The global population is increasing rapidly, with falling death rates, longer life expectancy and high fertility rates contributing to high populations of young people. This global growth is unequal, with the developed world largely experiencing declining population growth – partly due to low fertility rates – while in the developing world, still-high fertility rates mean rapid population growth.

In 2011, world population reached 7 billion, with most of this growth occurring in developing countries, especially in the very poorest. Ninety-seven per cent of this growth was in developing countries due primarily to still-high birth rates and young populations (high numbers in the reproductive ages).

'Almost all growth will take place in the less developed regions: today's 5.3 billion population in underdeveloped countries is expected to increase to 7.8 billion in 2050.'

www.populationmatters.org – Current population trends

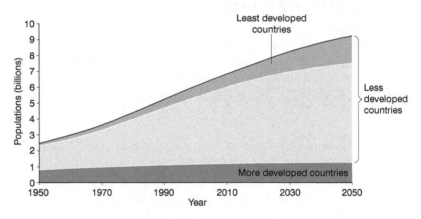

Figure 2.9.1 *World population prospects to 2050*

Source: United Nations Population Division, World Population Prospects: The 2010 Revision, medium variant (2011).

Demographic statistics – Latin America/Caribbean

Population growth rate	1.1%
% of total population of women aged 15–49	53
Total Fertility Rate	2.2
Life expectancy at birth	74
Population under 15 (%)	28
Population over 60 (%)	10

Source: www.unfpa.org – Country Profiles

Developing countries

Latin America and the Caribbean

This region has the smallest growth, due to fertility declines, and has the lowest average fertility rate of 2.2, closest to that of developed countries. Contraceptive use is very high in this region, but figures are also strongly influenced by the large countries of Brazil and Mexico. Two-thirds of married women use some form of modern contraception, but in Haiti only 25 per cent use modern contraception. The region's total fertility rate is 2.2.

Declining fertility means population growth in the region will be low but the population will rise to approximately 730 million by 2050 (an increase of 25 per cent). The region is also the most urbanised, with 77 per cent of the population living in urban areas. In addition, there is a high youth bulge, with a large proportion of the population concentrated in working ages rather than the economically dependent age groups.

Asia

There is a population of 4.3 billion in Asia, with China and India having responsibility for the most growth in the region (together they account for 60 per cent of the Asian population). India's current population of 1.3 billion is expected to exceed that of China (1.5 billion) by 2025 as its total fertility rate is very high. Within India, there are differences between the northern and southern states due to the disparities in family planning policies implemented by the states. Many other Asian countries have fertility rates of 1.4 or less (South Korea, Singapore, Taiwan). There is a wide range of TFR up to 6.4 (Afghanistan). Annual population growth rate is on order of 1.1 per cent. Whereas in China contraceptive use is over 90 per cent, outside of China, more than 50 per cent use contraception. More than 50 per cent of the world's 1.8 billion young people between 10 and 24 years live in Asia.

Demographic statistics – Asia

Population growth rate	1.1%
% of total population of women aged 15–49	55%
Total Fertility Rate	2.3
Life expectancy at birth	69
Population under 15 years	27
Population over 60	9

Source: www.unfpa.org – Country Profiles

CASE STUDY

Family planning and population growth in Uganda

With a population of 34 million and an average fertility of more than six children per woman, Uganda is one of the fastest-growing countries in the world. The reasons for the high birth rate range from strongly held religious beliefs, the practices of polygamy and early marriage, to the belief that large families are a source of prestige and wealth, coupled with the idea of child labour being an important source of income.

One 37-year-old woman already has seven children (eight in total – one died shortly after birth); she is currently using a contraceptive implant which is effective for three years. The high birth rate is also due to major gender inequalities, where men are unwilling to allow their wives access to contraception out of fear that their manhood would be diminished. In addition, 40 per cent of the health care in Uganda comes from faith-based organisations. Catholic hospitals and clinics do not stock any contraceptives. The problem is compounded by the poor road network throughout the country, as well as an unreliable distribution and supply.

Key points

- Global population continues to grow despite falling birth rates in some countries.
- Developing countries are responsible for more growth than developed countries.
- High fertility rates are largely responsible for rapid population growth.

Did you know?

In an effort to increase the population growth rate, the governor of Ulyanovsk region in Russia offers prizes to couples who have babies on Russia's national day, 12 June.

In developed countries, population growth is slowed or virtually nonexistent due to low birth rates and aging populations. In time, deaths will most likely exceed births in these countries.

Most developed countries have very low total fertility rates (TFRs), on the order of 1.3 or 1.4, and as a result they are facing severe population decline. Population growth in the developed world has been very low when compared to that in developing countries. Europe's population is projected to decrease from 740 million to 732 million by 2050. However, the population in the United States is expected to increase slightly due to its higher TFR of 1.9 in 2010.

Some European countries, notably France and Norway, have implemented pro-natalist policies to increase population growth in those countries. They have, for example, provided subsidised childcare, and generous maternal and paternal leave to help encourage larger families. Their TFRs are therefore slightly higher than other developed countries, standing at 2.07 in France and 1.97 in Norway.

Fertility rates have fallen sharply in developed countries, as women delay marriage and childbearing. Smaller family sizes are the norm, and contraceptive use is widespread. The Population Reference Bureau (PRB) makes mention of a demographic divide, which compares the situation in developed versus developing countries. On one side are mostly poor countries with relatively high birth rates and low life expectancies, while on the other are found mostly wealthy countries with birth rates so low that population decline is all but guaranteed. Life expectancy is quite high, extending past age 75, creating as a result rapidly aging populations.

Table 1 compares two countries, Tanzania and Spain, illustrating quite clearly the size of this demographic divide between developed and developing countries.

Table 1: The demographic divide

	Tanzania	Spain
Population (2012)	48 million	46 million
Projected Population (2050)	138 million	48 million
Lifetime Births per Woman	5.4	1.4
Annual Births	1.9 million	483,000
Percent of Population Below Age 15	45%	15%
Percent of Population Ages 65+	3%	17%
Percent of Population Ages 65+ (2050)	4%	33%
Life Expectancy at Birth	57 years	82 years
Infant Mortality Rate (per 1000 live births)	51	3.2
Annual Number of Infant Deaths	98000	1600
Percent of Adults Ages 15–49 With HIV/AIDS	5.6%	0.4%

Source: Population Reference Bureau

A comparison can also be made of developed countries themselves to show the characteristics which are common to this group of countries:

Table 2: Demographics of developed countries

Country	TFR	Population growth rate	Population under 15 years (%)	Population over 65 years (%)	Life expectancy at birth
Australia	1.88	1.5	19	14	82
Canada	1.67	1.1	16	14	81
France	2.07	0.5	19	17	82
Germany	1.37	0.0	13	21	80
Norway	1.97	1.3	19	15	81
Sweden	1.98	0.8	17	19	82
United Kingdom	1.95	0.6	18	17	80
United States	2.08	0.8	20	13	78

Source: Population Reference Bureau

CASE STUDY

In 2008, the EU's population stood at 495 million and was projected to rise to more than 520 million by 2035, before falling to 505 million by 2060.

'From 2015 onwards deaths would outnumber births, and population growth due to natural increase, would cease', said a population survey carried out by Eurostat, assuming a net migration inflow to the EU of almost 60 million over the next 50 years. 'Positive net migration would be the only population growth factor. However, from 2035 this positive net migration would no longer counterbalance the negative natural change.'

Across the EU's 27 countries there are now four people of working age for every person over 65, but by 2060 that ratio will be 2:1, causing stress on welfare and pension systems. Immigration is singled out as the sole mitigating factor, seen as crucial to maintaining population growth. But the report said this probably will not be enough to reverse the trend of population decline in many countries.

Fourteen of the 27 countries were projected to have smaller populations in 50 years' time; these included Bulgaria, Latvia, Romania and Lithuania.

The Guardian

Exam question

Examine Table 1 above, which gives demographic information for two countries: Tanzania and Spain.

1 Define the terms: **a** life expectancy   **b** infant mortality.   4 marks
2 Account for the difference in life expectancy in the two countries. 4 marks
3 Explain why lifetime births per woman is a valuable statistic for demographers.   4 marks
4 Discuss the consequences of the given population structure for future economic growth in Tanzania and Spain.   8 marks

Key points

- Developed countries have declining or shrinking populations.
- Incentives have to be given to increase population growth rates.
- Developed countries have very low fertility rates, often below replacement fertility rates.

✓ *Exam tip*

Even though you must answer all questions, read the entire question before attempting to answer it.

This helps to keep your answer coherent so that you do not waste time. It also helps to keep your answer on track. The question may ask for different aspects of a response in multiple sections; reading the question through will help you to organise your answer so it is given in the appropriate sections.

On completion of this section, you should be able to:

- recognise the cultural and social factors influencing population growth

- describe how population growth is influenced by cultural factors

- explain the relationship between culture and population growth.

Did you know?

In some developing countries, the cultural preference is for a male child to carry on the family name, or to look after their elderly parents. This can have an impact on population growth when parents keep having children in hopes of having a male child. It also has an impact on female births, many of which may end in abortion where selective screening has been carried out, using ultrasound tests to determine the baby's sex before birth; or in infanticide, when female babies are killed after birth. This has implications for future fertility rates.

Culture

Culture refers to a society's norms and values, and it often relates to aspects such as common use of language, behaviours and practices.

Age of marriage is one such cultural aspect of society. In societies where early marriage is the norm, the potential for population growth is high, since potential childbearing years are extended in a state of marriage. If a girl gets married at 16 years, as is the legal age of marriage in the Indian state of Rajasthan, she is likely to start having children right away, and over a reproductive lifetime of nearly 30 years she may have many children. Girls who marry early are also less likely to complete any formal education and therefore may have more children. The legal age of marriage notwithstanding, child marriage is still common in rural India, where girls as young as 12 years are married.

The way in which children are regarded is also part of the culture. When children are seen as the wealth of a family or society, each child is welcomed no matter the hardship. There is no limit set on the number of children a family may wish to have. In more affluent societies, however, children are often regarded in terms of their overall cost to the household, and most families may opt to limit the number of children they have. Additionally, in many developing countries, children contribute to the household through their labour in the fields or on the streets, with children as young as five engaging in some form of economic activity such as carrying stones to help build a road.

Large families may be held up as examples of great male virility or female fertility in some cultures, and the practice of having large families is therefore intertwined with the sense of self held by many in the population. Where polygamy is practised, there may also be very large families as one man may have many wives, and many children with each wife.

Religion

The role of religion in population growth sometimes focuses on the more stringent policies of some religions. In developing countries, government policies usually fall in line with the dictates of the church, despite the separation of church and state. In primarily Roman Catholic countries (for example, the Philippines), family planning programmes (especially the use of contraception) are not widely available, if at all, and abortions are illegal in all cases. This can lead to large families, especially among the poor, who may be more devout and likely to follow the strictures of the church. The stance of the Church on premarital sex may help curb teenage pregnancy.

Level and cost of education

The higher the level of education, the smaller families are likely to be. This is especially true of girls. When they receive more education, their families are healthier, and they tend to have fewer children. This is partly due to the fact that furthering their education occupies much of their childbearing years, so reducing the population growth, as well as the fact that a better education allows for more options other than marriage and childbearing. Many girls avoid marriage or put it off until much later, both of which reduce the overall population growth rate.

Social and economic status of women

As women gain social and economic freedoms, they are no longer restricted in having a say over their own reproductive health. Many women who are not independent earners, but dependent on men for their support, may be forced to have children to show their commitment or love. Their reproductive decisions may be taken away from them and they may be unable to even make the choice as to when or how many children to have. In Kenya, women have found ways around these strictures by attending family health clinics to have their children immunised, but where they are also educated about the use of contraception. Some women have the family planning injection, which is effective for up to three months at a time. In societies where women are seen as equal and treated as such, there is less need for family planning subterfuge, as women are able to take control of their own reproduction, deciding when and how many children to have.

 Exam tips

When you are asked to justify your response to a question, it is important that you state your reasons for the position which you are supporting. You are saying why your answer is correct. In these types of questions there may not be a clear-cut 'right' or 'wrong' answer. Be clear, be reasoned and be logical in your response.

CASE STUDY

The Philippines

Yolanda Naz and her husband are low wage earners in the Philippines, a country with a total population of 96.5 million, fertility rate of 3.27 and a median age of 22.2 years. Yolanda sells shampoo packets to her equally impoverished neighbours while her husband sells coconut drinks. They agreed early in their marriage to stop at three children, and although a devout Catholic, Yolanda took birth control pills.

After her third child was born, the mayor of Manila stopped the distribution of contraceptives at public clinics in order to promote a culture of life. Contraceptives became too expensive for millions of poor Filipinos to buy at private pharmacies. At age 36, Yolanda has eight children, most of whom are undernourished and often sick.

A reproductive health bill has been stalled by the Church for the past 14 years. This bill seeks to offer public education about contraception, and offer government subsidies to make them widely available. Women's control over their reproduction

is therefore being affected by religion and politics.

However, despite a devout population – 8 out of 10 Filipinos are Catholic – 70 per cent of the population supports this bill. In a 2008 survey, 35 per cent of married Filipino women in their childbearing years said that they wanted to avoid or postpone pregnancies but were not using modern contraception due to concerns about side effects, as well as husbands' opposition, the high costs and availability.

The former mayor of Manila, who removed contraception from public clinics, sees a growing population as economic potential. More people means a bigger labour force, bigger social security base, more productivity, more consumption and more production. He also believes that contraception weakens families and does not believe that governments should fund contraception.

Population in the Philippines is set to increase by more than half to 155 million by 2050.

Los Angeles Times

Key points

- Cultural, social and political factors all affect population growth rates.
- No one policy is wholly responsible for population growth in a country.
- Many country policies on population growth stem from cultural or religious practices.
- Educating girls has a great impact on reducing population growth rates.

Learning outcomes

On completion of this section, you should be able to:

- recognise the economic factors influencing population growth

- describe how population growth is influenced by economic factors

- explain the relationship between the economy and population growth.

Availability of pension schemes

In many poor developing countries where funds are limited, the availability of social security and pension schemes is limited. Many people in rural areas have no access to funds once they have retired, beyond what they are able to save over their lifetime. There is therefore a strong reliance on their children to look after them in their old age. Where health care may be limited and sanitary conditions lacking (such as a lack of running water), infant and child mortality rates can be quite high. Families are therefore large to ensure that some children do survive to adulthood and are therefore able to support their parents at this time. Where governments are able to look after the country's senior citizens, family sizes are low.

Level of affluence

It costs more to raise children in an affluent society than in a poor one. In developing countries, children are assets to their families, providing extra pairs of hands to work on the farm, or having the ability to generate their own income from early on. They are not a drain on resources. Many children do not go to school regularly and are therefore able to work in the fields or on the streets. In developed countries, children remain at home for long periods to be supported by their parents and attend school. Therefore, although in theory an affluent population can afford to have more children, they do not, due to the associated costs of raising children.

Economic development

As countries change from mainly agrarian economies to more service-oriented economies and beyond, there are changes in beliefs and behaviours with respect to population growth. Advances in health care and sanitation result in improved lifespans and lower child mortality. Family sizes begin to fall as more children survive infancy and childhood. Often, along with economic development there is social development, with increased opportunities outside of the home for girls and women, who then focus on their education and, increasingly, their careers. Marriages tend to be delayed as young people spend more time acquiring education, skills and qualifications. Therefore, fertility rates tend to also fall with increasing development. This pattern is seen throughout Europe as the population greys, or ages. This of course has repercussions for future population growth rates and economic productivity in the country.

CASE STUDY

Population change in the Caribbean

In some countries in the region, the population structure has reflected the movement in general towards greater levels of development. Barbados, the most highly ranked Caribbean nation (38) on the Human Development Index 2013, has 17 per cent of its population below the age of 15, with double that value aged 60 and over. In 2010, the median age was 37.5 years. This means that half the population was aged below 37.5 and half was aged above 37.5. The higher the median age, the older the population and usually the slower the population growth as persons move out of the reproductive age groups. This can have an impact on future economic growth as workers retire and there is no one to replace them.

Compared to this, in Haiti (ranked 161), 35 per cent of the population was aged below 15, with 13 per cent aged above 60 years. The median age was 21.5 in 2010; this reflects a highly youthful population and the potential for future growth is quite high as large numbers move into the reproductive ages. The impact of a youthful population on the economy and society can be quite severe. A young population means a high dependency ratio; the young people not of working age must be supported by those who are working. The provision of social services such as schools is also quite critical. In Haiti, the dependency ratio was 65.5 per 100 of the population aged 15–64, compared to 40 per 100 in Barbados.

It must also be recognised that an aging population would also require some assistance and support from the working population. Japan, ranked 10 on the Human Development Index, has a high dependency ratio of 59.6 per 100. It also has a very high median age (44.7), which is one of the world's highest.

Compounding the issue is the fertility rate. In Barbados, its fertility rate of 1.6 is below the replacement rate, and therefore leading to issues such as declining population and aging societies; in Haiti, a fertility rate of 3.2 means a rapidly growing population, as replacement rate has been met and surpassed. Japan's population would continue to age, with a fertility rate of 1.4.

Did you know?

Every second, five people are born and two people die: a net gain of three people. At this rate, the world population will double every 40 years and would be 12 billion in 40 years, 24 billion in 80 years, and more than 48 billion in 120 years. However, the United Nations estimates that world population will stabilise at 12 billion in 120 years, citing that effective family planning will result in a universally low birth rate. Education plays a key role: almost half of the world's population is under the age of 25.

Look here for more:
didyouknow.org/population

Key points

- Developing countries face greater economic challenges in reducing population growth rates.
- Provision of pensions and welfare for elderly people reduces population growth.
- Population growth in developed countries is very low as compared to developing countries.

On completion of this section, you should be able to:

- understand the term 'population control'

- understand the term 'direct population control'

- distinguish between pro-natalist and anti-natalist population control measures

- assess the effectiveness of population control measures.

Anti-natalist policies: measures and methods utilised by a country which are designed to reduce population growth.

Figure 2.13.1 Chinese one-child family

Population control

Population control refers to policies and strategies which are designed to limit or increase natural increase in a population. They are usually set by a government, and may be pro- or anti-natalist in scope. Some population control measures are direct, and designed to have a specific effect on fertility rates. Others deal with changes in thinking, which have an indirect impact on fertility rates in a country.

In order to maintain economic growth strategies and promote effective social development, many countries engage in some form of population control. This may be rigidly enforced policies and/or laws set out by governments, or more relaxed options available to the population based on their dominant cultural and religious beliefs. Few societies today are entirely self-regulating with respect to population growth control.

Direct population control

Anti-natalist policies

Direct population control includes the policies which governments enforce or endorse in order to increase *or* decrease the population growth rate. The most well-known example of direct population control is that of China's anti-natalist one child policy which was implemented to halt the country's burgeoning population growth in the late 1970s. However, while the population growth was slowed, there were some cultural and social issues, including a large gender gap between males and females. In some provinces in China, there are 120 males to every 100 males, which has severe implications for future population growth. A higher dependency ratio is also likely as the population ages without replacement.

Other countries have relied on other direct methods of population control. India has adopted varying strategies ranging from education about birth control, the establishment of family planning centres and clinics, and the education of girls, as well as sterilisation of both men and women after the birth of at least two children. In the Indian state of Kerala, the fertility rate has fallen to 1.8. This was achieved by decreasing mortality rates, reducing fertility rates and improving the health of young children. An important component was the emancipation of women, improving their role and status in society, and allowing them to achieve a greater measure of economic independence. This came about through increased education opportunities.

Pro-natalist policies

Pro-natalist societies *encourage* the growth of populations, usually by providing incentives. These policies are designed to counteract the falling fertility rates common throughout European countries. A decline in fertility coupled with an increase in longevity (greying society) may result in a decrease in labour supply, as well as the long-term prospect of a decrease in overall population size.

France, for instance, provides cash incentives to mothers to remain at home after the birth of a third child, reductions in the costs of train fares as well as income tax benefits, plus generous parental leave and subsidised day-care, and preferential treatment with respect to council housing for families with three or more children.

The establishment of a minimum age of marriage is another way in which population growth may be managed. When early marriages are common, many more children may be born than if marriages are delayed. Governments often change this as needed to either increase or decrease population growth rates.

Other direct methods of population control include incentive-based strategies such as:

- providing full legal rights to women
- increasing legal age at marriage for women
- providing payments for not having children
- priorities in jobs, housing, education for small families
- community improvements for achievement of low birth rate.

Disincentives include:

- higher taxes for each additional child
- higher maternity and educational costs for each additional child ('user fees').

Pro-natalist policies are not very common in the Caribbean, as most countries have been trying to reduce their population growth rates given the resource limits imposed by their small geographical size. Since the 1950s, most Caribbean countries have passed through Stage 1 and 2 of the Demographic Transition Model, and there are a number of countries where the fertility rate has fallen below replacement level of approximately 2.1, including Barbados (1.55), Trinidad (1.6) and Cuba (1.74).

Source of data: www.indexmundi.com

Caribbean countries are therefore on track for slow population growth. High levels of education, especially for women, are proving to be effective ways of reducing population growth. In Barbados, for example, the enrolment rate of women at the University of the West Indies or in other tertiary institutions is higher than that of males. In 2011 the ratio of women to men enrolled in tertiary education was 246 per cent, according to the World Bank (www.worldbank.org).

Key points

- Direct population control policies are usually the result of government policies.
- Pro-natalist policies support and encourage population growth.
- Anti-natalist policies try to reduce or stop population growth.
- Addressing other social issues such as the status of women is one way to slow down population growth.
- Coercive or punitive strategies have had limited success with population growth reduction, except in China.
- Most Caribbean countries are experiencing slow rates of population growth.

Did you know?

At one stage in Iran under the leadership of the Ayatollah Khomeini the age of marriage was lowered to nine. It was later raised to 15 after the population grew too rapidly and was too large for the available resources.

Pro-natalist policies: measures put in place by a government to encourage the growth of the population.

On completion of this section, you should be able to:

- understand the term 'indirect population control'
- assess the effectiveness of population control measures.

Indirect population control

The role of natural disasters (floods, hurricanes, earthquakes, volcanoes) in affecting population growth cannot be overlooked. In highly populated areas, the impact of natural disasters can be devastating for future population growth. If sufficient numbers of people in the reproductive age groups of 15 to 45 are killed or severely injured, there would be few offspring from that group.

In addition to the immediate deaths of people in this group, the dislocation and loss of secure homes and livelihoods can also have a negative effect on the population growth rate. Following a major natural disaster, it can take weeks or months to return to some state of normalcy, during which time issues such as a lack of housing and privacy, or the need to flee (becoming environmental refugees) may also cause a dip in population growth rates.

Haiti

In Haiti after the devastating earthquake of January 2010, population growth slowed. More than 200,000 Haitians died in the earthquake, a total nearly the size of Barbados' entire population. The associated health crises due to the spread of cholera also killed a large number of Haitians. The immediate displacement of thousands of people would also have contributed to the drop in population growth at that time as no one was living in settled accommodation, and people were constantly moving around, looking for safe places to live.

Social policies

Changing social policy is another indirect method of population control. Indirect population control aims to change the thinking about family size; to bring about a re-examination of the need or desire for larger or smaller families. Indirect policies try to change the way in which men and women relate to each other so as to allow women a greater say over their own fertility. This is often achieved through education, which gives women greater opportunities for work outside the home, or more knowledge about family planning and child health. This approach also tends to phrase population control in the guise of economic development policy, and so puts the focus on improving maternal and child health through family planning, rather than reducing population growth rates.

Other indirect policies to reduce family size:

- Promotion of secular education – *separation of religion and education.*
- Promotion of communication between spouses – *more likely to be discussion about desired family sizes and therefore some form of family planning.*
- Increased educational opportunities for women – *making more options available to women other than marriage and childbearing.*
- Lower infant and child mortality rates – *reduces the need for large families as insurance for elderly parents.*

www.jayhanson.us – How To Influence Fertility: The Experience So Far (1990)

Jamaica's population is currently estimated at 2.65 million, of which women in the reproductive age cohort of 15–49 years represent approximately 707,600, or 26.7 per cent. The population growth rate has been declining steadily since 1995, and is currently at a rate of 0.5 per cent. Jamaica's current fertility rate of 2.34 has raised concerns for local government, and they are undertaking a 5 year plan to reduce it to 2.2. This will be achieved mainly by:

■ increasing the availability of contraception over the island

■ training health care providers and counsellors to aid in the process of providing contraception and support

■ increasing education especially among young people through the use of public service announcements

■ promoting abstinence in young people by using peer pressure.

The Jamaica Gleaner

Indirect population control can cause population growth or decline. The overall decline in population is often the result of the effects of natural disasters, or simply the economic and social policies in place, such as high taxation, or transport costs. Population control may simply be a by-product of the policies or the occurrences, rather than the desired result.

'In India, in 1994 a 'New Population Plan' (NPP), was implemented in the hope that that by the year 2010, the average total fertility rate would fall from 3.4 to around the replacement rate of 2.1. In order to achieve this goal, the NPP allowed universal access to contraceptives and promoted greater education on contraception, trained more people to aid in the birth of children, required a formal registration of all marriages and births, maintained and enforced the minimum age of marriage at 18, and aimed to provide primary education for more citizens.'

www.colby.edu

An increase in population may result from the enhanced provision of services not directly related to population; for example, improvements in the rights of women in society especially with respect to job security may result in a rise in population growth rates as women are assured that they will not be 'penalised' for taking time off to have children.

Governments do engage in both types of population control over time as population growth rates fluctuate, as has been the case in Iran. In the Caribbean, however, population reduction rather than growth has been the general trend.

☑ Exam tip

When you are asked to discuss a concept, question, or statement it is important that you give more than one point of view in your response. Even if you agree with the statement presented, a full discussion will also address opposing views.

Key points

■ Indirect population control results from policies which do not specifically target population growth.

■ Improved access to education is one indirect way of reducing population growth.

■ Indirect population control may result in increased *or* decreased population growth rates.

■ Many countries have made use of both indirect and direct population control strategies at different times in their history.

Measures of poverty

Access to education

There are many factors influencing one's access to education, these include:

Economic factors

- Availability of educational services – universities, colleges, tertiary institutions
- Financial assistance, e.g. GATE in Trinidad and Tobago, Student Revolving Loan Fund in Barbados
- Government policies, e.g. universal childhood education, ICT infusion, Laptop initiative in Trinidad and Tobago, Guyana.

Socioeconomic factors

- Personal income/family income level
- Parents' level of education
- Ethnicity and gender
- Parental support
- Culture/religion.

Access to health care

'[Health is] the state of complete physical, mental and social well-being and not merely the absence of disease or infirmity' and the 'extent to which an individual or group is able to realise aspirations and satisfy needs, and to change or cope with the environment. Health is a resource for everyday life, not the objective of living; it is a positive concept, emphasising social and personal resources as well as physical capabilities'

Source WHO, 1948 and 1984.

There are many factors influencing one's access to health care, but the most common health indicators are those relating to birth and death:

- Life expectancy
- Premature mortality (e.g., Years of Potential Life Lost or YPLLs)
- Cause of specific deaths (e.g., lung, cervical cancer)
- Access to and adequacy of prenatal and postnatal care.

Indicators	Examples
Morbidity/Health Status	- Obesity – Body Mass Index - Diseases – diabetes, asthma, and other chronic diseases.
Health behaviours	- Regular physical activity/exercise - Diet and nutrition - Lifestyle – smoking, drug and alcohol use.

Access to health care	■ Availability of health care facilities – public hospitals, private hospitals, medical centres
	■ Access to specialised services (e.g. cancer treatment centres – Brian Lara Cancer Treatment Centre of Trinidad and Tobago)
	■ Insurance coverage
	■ Receipt of preventive services.
Physical environment	■ Area-based measures, e.g. population density
	■ Access to public transportation, housing
	■ Environmental pollution, e.g. air and water quality.
Social environment	■ Income (individual/family)
	■ Education (e.g. high school graduation rates, basic literacy and numeracy levels
	■ Social support and connectedness.

Basic human needs

The traditional list of basic needs has been modernised over time to include other elements that a country might value, but these elements vary from one country to the next (love, air). Throughout the world the concept of family is changing and the need for emotional stability is key to a country's future productive capacity, hence the inclusion of love as a basic need. Education at all levels has become a necessity over time as countries recognise the importance of developing their human capital and the possible economic gains from such an investment.

According to Abraham Maslow, needs must be satisfied in a particular order starting from the lower-level needs before our higher-level needs can be fulfilled. Our basic needs lies in the band at the physiological level.

Absolute vs. relative poverty

Poverty is often defined in either relative or absolute terms. **Absolute poverty** measures poverty in relation to the amount of money that people need to meet the basic needs discussed above; people are described as being in absolute poverty particularly if they lack resources to obtain enough food to achieve the minimum calorie intake needed for basic good health. The concept of absolute poverty is somewhat limited when it comes to describing poverty, however, because it does not take account of the overall level of inequality in society, or people's needs beyond the most basic ones for survival. The concept of **relative poverty** does take into account these needs, because it defines poverty in relation to the economic status of other members of the society: people are poor if they fall below the minimum acceptable standard of living established in a given society or country.

Figure 2.15.1 *Components that make up the necessities of life*

∞ *Links*

data.worldbank.org/country
www.oecd.org

Key points

- Levels of poverty can be measured by indices, of which access to education, access to health care and access to basic needs are the most important.

- Absolute poverty occurs when people cannot obtain adequate resources (measured in terms of calories or nutrition) to support a minimum level of physical health.

- Relative poverty exists when people do not enjoy a minimum level of living standards as determined by a government.

Learning outcomes

On completion of this section, you should be able to:

- discuss the Human Development Index (HDI) and the Gender-related Development Index (GDI)

- calculate the GDP, GNP and HDI of a country.

Per capita income

Per capita refers to the income earned per person in a given population. Per capita income is often used to measure a country's standard of living, which can be calculated by dividing national income by population. Different measures of national income like GDP and GNP can be used and accordingly we get per capita GDP, per capita GNP and so on. If we use real national income it is called real per capita income.

Gross Domestic Product (GDP)

The definition of GDP is the monetary value of all the finished goods and services produced within a country's borders in a specific time period, though GDP is usually calculated on an annual basis. It includes all private and public consumption, government outlays, investments and exports less imports that occur within a defined territory. It is also considered the sum of value added at every stage of production of all final goods and services, within a country over a given period of time.

When an economy achieves full employment of all resources/production factors, indicated by a very low unemployment rate, it is producing at its **potential GDP**.

The most common approach to calculating GDP is by expenditure:

GDP = (C + G + I) + NX

or

GDP = (C + G + I) + X – M

where:

C is equal to all private consumption, or consumer spending, in a nation's economy

G is the sum of government spending

I is the sum of all the country's businesses spending on capital

NX is the nation's total net exports, calculated as total exports minus total imports.

(NX = Exports – Imports)

M is the total import

X is the total export

Gross National Product (GNP)

GNP is a measure of a country's economic performance, or what its citizens produced (i.e. goods and services) and whether they produced these items within its borders. It is calculated using the GDP of a country, plus any income earned by residents from overseas investments, minus income earned within the domestic economy by overseas residents. GNI per capita (formerly GNP per capita) is the gross national income, converted to US dollars using the World Bank Atlas method, divided by the midyear population.

Human Development Index (HDI)

The Human Development Index (HDI) is a summary measure of a country's average achievements in three basic dimensions of human development: a long and healthy life (Health/Life Expectancy), access to knowledge (Education) and a decent standard of living (Income).

The index was created by economist Mahbub ul Haq, and further developed by economist Amartya Sen in 1990. However, in its 2010 Human Development Report, the United Nations Development Programme (UNDP) began using a new method of calculating the HDI. The following three indices are used:

1 Life Expectancy Index (LEI) 2.

2 Education Index (EI) – Mean Years of Schooling Index (MYSI) and Expected Years of Schooling Index (EYSI).

3 Income Index (II).

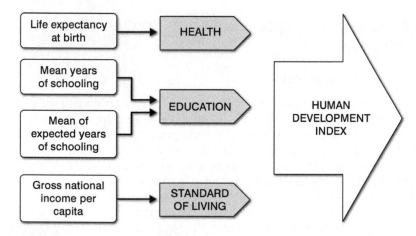

Figure 2.16.1 *Components of the HDI*

Did you know?

The HDI sets a minimum and a maximum for each dimension, called goalposts, and then shows where each country stands in relation to these goalposts, expressed as a value between 0 and 1.

Gender-related Development Index (GDI)

This index measures achievement in the same basic capabilities as the HDI does, but takes note of inequality in achievement between women and men.

The methodology used imposes a penalty for inequality, such that the GDI falls when the achievement levels of both women and men in a country go down or when the disparity between their achievements increases. The greater the gender disparity in basic capabilities, the lower a country's GDI compared with its HDI. The GDI is simply the HDI discounted, or adjusted downwards, for gender inequality (UNDP).

 Links

More useful information on this topic can be found here:
hdr.undp.org/en

 Exam tip

Know the top 10 countries with the highest HDI and provide details of why they have such a high ranking.

Key points

- While it is not necessary to know how to calculate all the indices, you must be able to write the formula for the calculation of GDP.

- The GDP and GNP are economic indicators of a country's performance while the HDI and GDI are holistic indicators of development of a country reflecting several indices.

- Per capita measurement is useful to compare consumption patterns of resources, both locally or with other countries.

Learning outcomes

On completion of this section, you should be able to:

- briefly state the relationship between population growth and poverty
- describe the environmental impacts of population growth.

Population growth and poverty

Unchecked growth in population will continue to place substantial stress on the capacity of the environment to provide. As population numbers continue to rise in developing countries and within low-income communities, due primarily to the lack of jobs, access to and knowledge about family planning strategies, slow per capita income growth, a lack of progress in reducing income inequality, and more poverty. Conversely, many characteristics of poverty, such as high infant mortality, lack of education for women, and inaccessibility of family planning, can cause high fertility – perpetuating the cycle of poverty.

Environmental impacts of population growth in the Caribbean

Deforestation in Haiti

In Haiti deforestation is both a symptom of overpopulation and the root cause of human suffering. Haiti is an example of what happens when too many people have too few natural resources in a limited environment. The primary cause of deforestation in Haiti was the need for a cheap alternative energy source and pasture lands for livestock. Only 10 per cent of Haiti's population has access to electricity. Unfortunately for Haiti's natural environment, wood became and continues to be the principal energy source, accounting for 70 per cent of the country's energy consumption. This resulted in the steady removal of trees with an estimated 6000 hectares of soil lost each year to erosion. The use of coal as an energy source ensures revenue earning by locals who traditionally sell charcoal by the roadside. The lush green mountains of Haiti have been replaced by areas completely devoid of vegetation, bare mountains and advanced soil erosion on land that continues to support a growing population. Today 98 per cent of the trees on Haitian soil have been cut, ranking it among the five worst environmental disasters of the world.

Based on an article in The Foreign Policy magazine, 2010

Figure 2.17.1 Deforestation and its impact on Haiti's landscape

Figure 2.17.2 Roadside sale of coal in Haiti

Pollution in the Caribbean Sea

According to the National Center for Ecological Analysis and Synthesis (NCEAS) in the USA, the Caribbean Sea is the second largest most polluted sea in the world. The deterioration of these Caribbean coastlines has been attributed to three major human activities: overfishing, climate change and tourism – all in response to an increase in population. The use of new types of catching gear and destructive fishing practices has led to drastic changes in fish stocks as well as endangering the region's marine food web. The shortage of seafood and the depletion of the fishermen's source of income is just one economic side effect of the Caribbean Sea's degradation.

Did you know?

The Caribbean is the second largest sea in the world and covers an area of more than 3.2 million square kilometres.

Environmental deterioration in Jamaica

Tourism poses a major threat to coastal environments through development, resource extraction and overuse. Tourists are often drawn to the very same areas that are often fundamentally important to the livelihood and enjoyment of local people. While tourists and locals compete for the same resources, it is the locals who are negatively affected. As population pressures increase coastal management suffers, leading to ecologically unsustainable practices and other sectors utilising these environments also suffer. Fisheries and other resource extraction generally stresses coastal and marine ecosystems, limiting productivity and ruining land- and seascapes. Tourism can harm wildlife endemic to the Caribbean islands via the introduction of alien invasive species as they compete for habitat space, disrupting food chains or introducing foreign disease. Jamaica currently has the highest number of threatened animal and plant species in the Caribbean at 254. The small Indian mongoose is believed to have contributed to the possible extinction of two of Jamaica's endemic birds: the Jamaica petrel and Jamaican pauraque. It is also a threat to the endemic Jamaican iguana.

Climate change in the Caribbean

Scorching temperatures, sea-level rise and increased hurricane intensity and frequency threaten lives, property and livelihoods throughout the Caribbean region. As ocean levels rise, these low-lying islands will experience coastal submergence. As temperatures rise and storms become more severe, the aesthetic environment deteriorates, negatively impacting on tourism – the revenue earner of many Caribbean economies. And these devastating impacts will occur regardless of the fact that Caribbean nations have contributed little to the release of the greenhouse gases that drive climate change.

Did you know?

The principal causes of environmental degradation are twofold: in developing countries the issue is population growth but in the developed region the issue is the over-consumption of resources.

Key points

- Population numbers affect more than just the environment, but the environment captures attention because of its role in providing the key elements for sustaining human life.

- Regulating population is the key when addressing issues because people require and demand resources: land, water, air, food.

On completion of this section, you should be able to:

- describe current geographical variation in human consumption patterns

- compare consumption patterns in the Caribbean

- explain the impact of lifestyle on consumption patterns in the developing and developed countries.

Did you know?

1 One cubic metre of water is equivalent to 264.17 gallons.

2 Per capita water consumption is an indicator between demand and supply and the need to implement water resource management.

3 Hydrologists typically assess scarcity by looking at the population–water equation. An area is experiencing water stress when annual water supplies drop below 1700 m^3 per person. When annual water supplies drop below 1000 m^3 per person, the population faces water scarcity, and below 500 cubic metres 'absolute scarcity'.

Consumption patterns in the Caribbean

Urbanisation is the major driving force behind the rate and pattern of consumption of resources. This pattern of consumption varies significantly in urban and rural areas as there is a strong positive relationship between income and consumption of resources. Diets in developing countries have shifted away from the traditional staples, such as cereals, roots and tubers, to a variety of meat products, dairy and oil crops and sugary foods.

Water

The Caribbean is facing an ongoing water-shortage crisis, due primarily to the reckless high patterns of consumption. Although on track to meet the Millennium Development Goals regarding coverage in water distribution and sanitation, the quality and quantity of services is severely deficient. Many islands in the Caribbean experience disruptions in their water supply or they are placed on a water schedule in an attempt to control water use. A large volume of water is loss through leakages due to structural defects and a lack of routine maintenance. The average water consumption varies within the Caribbean. In Jamaica it is 224 m^3 per person per year; in Trinidad and Tobago the figure is slightly lower at 177 m^3/per person per year, but in Guyana it is 2161 m^3.

Food

Guyana is one of the few countries in the southern hemisphere that has a net export of food, achieving the United Nations first Millennium Development Goal of eradicating hunger through 'The Grow More Food Campaign' aimed at increasing food production. The campaign called for the implementation of a US$21.9 million Agricultural Export Diversification Programme, the implementation of a US$6 million Rural Enterprise and Agricultural Development Programme, and increased investment in drainage and irrigation by restoring drainage to areas abandoned by farmers, and training farmers to manage the maintenance of rehabilitated structures.

Caribbean countries are currently experiencing rapid dietary, or nutritional, and demographical transitions. A shift is taking place away from indigenous staples, local fruits, vegetables and legumes, to more varied energy-dense diets based on more processed foods/beverages, animal products, added sugars, fats and alcohol.

Purchasing ready-to-eat meals and eating out are widely prevalent in Barbados: 45.3 per cent of men and 31 per cent of women consume ready-to-eat meals from fast-food restaurants at least once or twice a week.

Fuel

Worldwide fuel consumption averages 1853 kilograms of oil equivalent per person per year. In the Caribbean the agricultural and industrial sector accounts for a larger portion as compared to the residential and service sector. The largest consumer of fuel in the Caribbean is Trinidad and Tobago, primarily because it has large reserves of natural gas and oil but also due to the increase in demand by the manufacturing and industrial sector and increased car ownership. Trinidad and Tobago consumes an average of 331 kg of oil equivalent as compared to Jamaica's 19 kg.

Greenhouse gas emission

The main greenhouse gases affecting our planet today are carbon dioxide, methane, and nitrous oxide and fluorinated gases. In the Caribbean region the drive towards development compels small islands like ours to utilise the natural resources of the region despite the environmental cost.

The major sources of these gases are:

Country	CO_2 (t/person/year)
Canada (all regions)	17.37
Dominican Republic (all regions)	1.98
Jamaica (all regions)	4.74
Trinidad and Tobago (all regions)	21.85
United Kingdom (all regions)	8.6
United States (all regions)	19.1

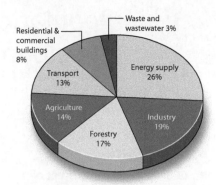

Figure 2.18.1 Global greenhouse emission by source

Source: IPCC (2007); Global Greenhouse Emission by source

Waste production

According to the Commission on Sustainable Development (CSD) Inter-sessional Conference on Building Partnerships for Moving towards Zero Waste, solid waste generation varies from 0.7 kg/person/day to 2.8 kg/person/day in Caribbean countries. Steady increases in population numbers, urbanisation, changing consumption patterns, trade and seasonal tourism all account for the changes in waste composition and the overall increase in waste production.

Country	Waste generated (kg/per person/day)
St Lucia	1.45
Grenada	0.85
Trinidad and Tobago	1.81
Jamaica	1.5
Barbados	0.9

Did you know?

Trinidad and Tobago generates the most waste per person per day, due primarily to its diverse economy whereas countries like Barbados and St Lucia have undertaken measures to reduce waste generation and waste disposal.

 Exam tip

It is important to know why there is a disparity in consumption patterns in developed versus developing countries.

Key points

- Consumption of resources is increasing annually.
- The rate of consumption is closely linked to the natural availability of that resource in the country.
- Tap water is often not potable, leading to growing reliance on bulk bottled water consumption.

On completion of this section, you should be able to:

- briefly explain the concept of sustainable development
- state the goals of sustainable development
- state key indicators of sustainable development
- explain the relationship between population growth and sustainable development.

Concept of sustainable development

The term 'sustainable development' was popularised in *Our Common Future*, a report published by the World Commission on Environment and Development in 1987, also known as the Brundtland report. It defined the concept of sustainable development as: **'development which meets the needs of the present without compromising the ability of future generations to meet their own needs'.**

True levels of sustainable development depend heavily on the ability of any country to:

1 Commit to equity and fairness – requires the development of policies geared towards the alleviation of poverty by satisfying basic human needs.

2 Understand the limitations imposed by the state of technology and social organisation on the environment's ability to meet present and future needs. The precautionary principle states that 'where there are threats of serious or irreversible damage; lack of full scientific certainty shall not be used as a reason for postponing cost-effective measures to prevent environmental degradation' (*Rio Declaration on Environment and Development*, Principle 15).

3 Sustainable development calls for decision making that reflects the convergence of the three pillars of economic development, social equity and environmental protection.

Goals of sustainable development

One of the main outcomes of the Rio+20 Conference on Sustainable Development held in 2012 was the agreement by member states to initiate the development of Sustainable Development Goals (SDGs) and key indicators that would track progress and lead to a solid outcome. The United Nations, in collaboration with the World Bank and the OECD, identified six goals for sustainable development.

Goals and indicators of sustainable development

	1 Poorest Fifth's Share of National Consumption
	2 Child Malnutrition: Prevalence of Underweight Under 5s
Environmental Sustainability	Environmental Sustainability
Poverty, Environment, and Natural Resource Practical implementation of sustainable development should be delivered at national and international level to fight poverty and protect the environment. (Johannesburg Summit 2002)	*Examples of Target from the Johannesburg Plan of Implementation*
	1 Poverty: Reduction in Population living on less than US$1 Per day.
	2 Water and Sanitation: People with Access to Safe Water and Basic Sanitation.
	3 Energy: Accessibility to Environmentally Sound Energy Supply.
	4 Health: Enhancement of Health Education.
	5 Sustainable development of small island developing States: Development of Community-based Initiatives on Sustainable Tourism.

Social Development	Social Development
Universal primary education There should be universal primary education in all countries by 2015. (Jomtien, Beijing, Copenhagen)	**1** Net Enrolment in Primary Education. **2** Completion of 4th Grade of Primary Education. **3** Literacy Rate of 15 to 24 Year-Olds.
Gender equality By eliminating gender disparity in primary and secondary education by 2005. (Cairo, Beijing, Copenhagen)	**1** Ratio of Girls to Boys in Primary & Secondary Education. **2** Ratio of Literate Females to Males (15 to 24 Year Olds).
Infant & child mortality The death rates should be reduced by two-thirds of the 1990 level by 2015. (Cairo)	**1** Infant Mortality Rate. **2** Under 5 Mortality Rate.
Maternal mortality Should be reduced by three-fourths between 1990 and 2015. (Cairo, Beijing)	**1** Maternal Mortality Ratio. **2** Births Attended by Skilled Health Personnel.
Reproductive health Reproductive health services through the primary health care system for all by 2015 (Cairo)	**1** Contraceptive Prevalence Rate. **2** HIV Prevalence in 15 to 24 Year Old Pregnant Women.

ESCAP

Population and sustainable development

Population size, per capita consumption and the environmental damage caused by the technology and processes used to produce what is consumed determines the overall human impact on the environment. Developed countries have the greatest impact on global environment conditions to date, as 20 per cent of the world's people living in the highest-income countries are responsible for 86 per cent of total private consumption compared with the poorest 20 per cent, who account for a mere 1.3 per cent of overall private consumption.

As population numbers continue to increase, the desire especially in the developing world to achieve a suitable standard of living is justifiably high. Any increase in population numbers will amplify the rate of consumption of resources, as there is greater demand on the ability of the natural environment to meet these changes in demands.

Key points

- Both current and new consumers worldwide need to realise and address the consequences of their levels of consumption.

 Links

sustainabledevelopment.un.org
www.unescap.org

Changes in consumption patterns

The current rate of consumption is undermining the existing environmental resource base; that is, the finite supply of water, air and land. The high consumption, high waste lifestyle of the top-earning fifth of the world's population is exacerbating inequalities worldwide. If this trend is allowed to continue, the right of the lowest-earning fifth of the world's population to satisfy their basic needs will be violated.

Changes in diets

Modernisation of the food pattern in Latin America and the Caribbean has been inspired by the US food pattern, which is dominated by fast-food services. The average US diet is rich in fats, refined sugar, additives, salt and poor in fibre. It is poor in complex carbohydrates and most of its protein is of animal origin. Although one recent consumption trend, healthy eating, tends towards organic products with less processing and an increase in use of fresh fruit and vegetables.

Urbanisation

Urbanisation and globalisation may enhance access to non-traditional foods as a result of changing prices and production practices, trade and marketing practices. These forces have influenced dietary patterns throughout the world. How we consume drives how we extract resources, create products and produce pollution and waste.

Individual preferences and beliefs

Our physiological needs provide the basic determinants of food choices but it is an individual palate that determines what is consumed. Palatability is proportional to the pleasure someone experiences when eating a particular food.

Cultural traditions

According to historian Donna Gabaccia 'food and language are the cultural habits humans learn first and the ones they change with the greatest reluctance'. The meaning of food for different cultural groups goes beyond providing sustenance. Cultural food patterns are defined by what, when, how, and with whom foods are eaten. Ethnic and racial groups differ in how they identify foods and how they prepare them, the condiments they use, and the timing and frequency of meals. Foods are frequently used in symbolic ways, playing an integral role in religious ceremonies and social events. Cultural food practices are dynamic and ever-changing, although many traditions persist with acculturation.

Income/commodity pricing

The cost of food is a primary determinant of food choice. Whether cost is prohibitive depends fundamentally on a person's income and socioeconomic status. Low-income groups have a greater tendency to consume unbalanced diets; however, access to more money does not automatically equate to a better quality diet but the range of foods from which one can choose increases.

Constraints on sustainable development

Strategic Imperative for Sustainable Development #4 'Ensuring a Sustainable Level of Population' (*Our Common Future*, **Brundtland Report, 1987**)

Increase in population numbers

Population growth rate is a key element affecting sustainable development. However, the issue is not necessarily the size of the world population but the location and the rate of consumption that follows. For example, a child that is born in a country where levels of material and energy use are high places a greater burden on the Earth's resources than a child born in a poorer country. Sustainable development can be pursued more easily when population size is stabilised at a level consistent with the productive capacity of the ecosystem.

Changing role of women in society

Fifty-one per cent of the world's population consists of women, and addressing their changing role is society is vital in reaching maximum economic production. Today women are challenging and educating themselves and are entering jobs in all sectors of the economy. As women advance their careers, many are opting to postpone family life, changing the existing population structure. Though both women and men work outside the home, women are still expected to fulfil both their domestic and work obligations. There is a need for greater equality between men and women in the home as well as the workplace to facilitate the advancement of women.

Access to basic amenities

Cities in developing countries are growing much faster than their capacity to cope, creating shortages of housing and water, adequate sanitation and mass transit facilities. A growing proportion of city-dwellers live in slums and shanty towns, and further deterioration is likely, given that most urban growth will take place in these urban centres, and this will compound the pressure on the existing resources and services.

Urbanisation

Urbanisation should be a planned process that is managed to minimise any negative impact in the quality of life experienced by the inhabitants. The extension of core services to growing towns and cities in an attempt to alleviate the demands on the capital. The renewed focus on the agricultural sector for food security, self-sufficiency and economic diversity can also serve to attract urban dwellers to more rural environments, as well as encouraging rural inhabitants to invest. Governments have developed financing and support systems to promote and encourage development in these unpopular and abandoned areas.

Key points

- The demand for resources will continue to increase as population growth takes place.

- Population growth and economic viability determine the rate of resource exploitation and depletion.

- Mass consumption and the growth of a materialistic society continues to extract resources in an unsustainable manner.

Did you know?

'Over'-population is often highlighted as the major cause of environmental degradation but consumption patterns today are not able to meet everyone's needs, therefore the system that drives these consumption patterns also contribute to inequality of consumption patterns.

☑ *Exam tip*

- Consumptive use of resources could have a negative impact on the environment but the non-consumptive use can be beneficial to both the environment and the economy as an income generator.

- Know the concept of Demand Management and how it impacts on resource use.

On completion of this section, you should be able to:

- explain the principal ways in which people impact negatively on the environment
- examine the environmental impacts of over-consumption in developed and developing countries.

Did you know?

The voracious global appetite for fish, of which Japan's huge demand is a key part, has led to widespread overfishing and pushed many highly sought after species, including some varieties of bluefin tuna, close to the edge. Japan exceeded its quota by 25 per cent in 2012, while France, Portugal and Spain have repeatedly overfished the same species in the same area in recent years as a result of their quota being significantly reduced.

Figure 2.21.1 *Unsustainable fishing practices*

Over-consumption of resources

Overexploitation: Fishing

Overexploitation is the unsustainable use of resources whereby the rate of extraction is beyond the regeneration capacity of the environment, leading ultimately to resource depletion. Current consumption patterns, international trade agreements and increases in populations have been identified as the driving forces behind the rate of resource extraction. According to the United Nations Food and Agriculture Organisation (FAO) over 25 per cent of the world's fish stocks have either been overexploited or depleted and 52 per cent are fully exploited. Thus a total of almost 80 per cent of the world's fisheries have been fully overexploited, depleted, or are in a state of collapse. The size of fishing fleets, the use of modern techniques to facilitate harvesting, transport and storage and the increase in demand have exceeded what the oceans can sustainably support.

The ease of access to a fishing licence and the establishment of fishing cooperatives led to the expansion and over-supply of fish resources in Belize. Overfishing and illegal fishing have depleted stocks of grouper and snapper, Caribbean spiny lobster and queen conch so that the numbers and size of the catch have diminished. Dwindling numbers have forced fishermen to focus on less popular species like parrotfish, doctorfish and the surgeon. According to Hon. Lisel Alamilla, Minister of Forestry, Fisheries and Sustainable Development, 'The greatest challenge that we face is overfishing in our seas'.

Habitat destruction

Scientists agree that species extinction is a direct result of habitat destruction. Habitat loss affects local and global biodiversity, especially where:

- the species are endemic to an area
- animal are considered as food specialists
- species are highly migratory
- species are lower down the food chain.

Land is a valuable commodity today. Competing uses include housing, cultivable land for agriculture, urban sprawl and industrial development. However, the underlying cause for land shortages is the continued growth in population numbers. The Amazon Basin is the focal point in habitat destruction, leading to an increase in the greenhouse effect, flooding, loss of plant and animal species and loss of soil fertility. A rapidly growing industry in America is that of self-storage. Thousands of acres of good farm land are paved over every year to build these cities of orphaned and unwanted things so as to give people more room to house the new things that they are persuaded to buy. If these stored products were so essential in the first place, why do they need to be warehoused?

Pollution

Whether rich or poor, people in every corner of the world are confronted with the pressing challenge of water pollution. While Jamaica is experiencing an overall decline in its physical state due to increases in pollutants, its water quality status is most pressing. Jamaica has 9.4 km^3 of renewable water resources, with 77 per cent used for agriculture and 7 per cent used for industrial purposes. About 85 per cent of the people living in rural areas and 98 per cent of the city dwellers have access to pure drinking water. Water supply sources and connected downstream wetland, estuary and marine ecosystems are being contaminated by sediments, pathogens and chemicals. This contamination results primarily from deforestation of upper watersheds, improper disposal of liquid and solid wastes, and runoff contaminated by agrochemicals. The principal consequences of this water pollution are noticed in human health impacts and degradation of downstream habitats. Inadequate planning of urban development and the expansion of numerous illegal housing and other developments is a major factor contributing to water pollution. Many of these structures, whether legal or illegal, do not have access to reliable piped water systems or waste disposal services.

Landfills swell with cheap discarded products that fail early and cannot be repaired. Planned obsolescence, also called built-in obsolescence, refers to the design and production of a product, such as hardware or software, with the intent that it should be useful, functional or popular for a limited length of time. A generation is growing up without knowing what quality goods are. Friendship, family ties and personal autonomy are often promoted as a vehicle for gift giving and the rationale for the selection of communication services and personal acquisition. Much becomes mediated through the spending of money on goods and services.

Introduction of exotic species

The Caribbean is one of the world's biodiversity hotspots. However, many of its native species are threatened by invasive alien species (IAS) from other countries, which continue to be brought in at an alarming rate by trade, transport, travel and tourism (the 'four "T"s'). Currently there are 552 alien species identified in the Caribbean region alone. IAS in the Caribbean region have caused serious problems to: (1) human health, for example multiple deaths caused by the H1N1 virus (2) biodiversity, for example decimation of native fauna species (reptilian) caused by the Indian mongoose (3) agriculture, through loss of production yields via unintentional introduction of the tropical bont tick, which has caused a decline in ruminant production. Recently, Trinidad and Tobago has experienced a surge in the frequency of introductions with respect to exotic species, such as the citrus black fly, citrus leaf miner, black Sigatoka disease, the red palm mite, coconut moth and the giant African snail. This coincides with the increase in the importation of agricultural commodities which has the greatest potential for the entry into a country.

Did you know?

Over-consumption is a direct response to the wants and needs of a population.

People overpopulation exists when the rate of resource consumption is small but, but there are so many people that the resource is still depleted.

Consumption overpopulation exists when the population is small compared to other regions, but each person uses so many units of a resource that the resource still becomes depleted.

Did you know?

Exotic species are also referred to as introduced, invasive alien, non-indigenous, pest or non-native species.

Key points

- The underlying cause of resource over-consumption and depletion is the increasing number of people demanding the same resources.

- Materialism and consumerism are key factors in the misuse of scarce but valuable resources.

- Consumerism is manifested economically in the chronic purchasing of new goods and services, with little attention to their true need, durability, product origin or the environmental consequences of manufacture and disposal.

On completion of this section, you should be able to:

- explain the mitigation measures taken to address the negative roles played by humans on the environment.

Mitigation measures

Changes in lifestyles

Lifestyle changes are necessary to alleviate the demands on the natural environment to supply an over-abundance of resources to meet the demands of an ever-increasing population.

Action	Mitigation
Overexploitation	Alter behaviour by creating seasonal extraction and a total ban on sensitive and endemic species.
	Encourage the consumption of other, more abundant species.
Habitat destruction	Recycling of forest products.
	Stop junk mail.
	Do not buy endangered animal products.
	Buy products that are recycled.
	Seasonal ban.
Pollution	Use CFC-free products.
	Boycott companies that produce CFCs.
	Carpool.
	Use energy efficient products.
Introduction of exotic species	Promote the use of local natural resources.

Use of substitutes

Action	Mitigation
Overexploitation	Establish a quota for resources extraction.
	Replace hardwood products with synthetic products to reduce the demand on forest resources.
Habitat destruction	Replacement of soft wood with more durable material like block and steel.
Pollution	Use solar-powered appliances in place of electrically powered devices, e.g. solar-powered water heaters.
Introduction of exotic species	Promote local species.

Did you know?

Substitute resources or environments have been used in some instances to mitigate unavoidable impacts. Substitutes can be effective in reducing the pressures on natural sources to keep pace with the ever-increasing demand.

Application of environmentally friendly technology

Action	Mitigation
Overexploitation of water sources	Conserving water resources by the use of decentralised systems. Promoting rainwater collection systems.
Habitat destruction	Using resources more efficiently by deriving maximum output per unit, e.g. wood to pulp to paper.
Pollution	Using wood reinforced with fibreglass and renewable energy.
Introduction of exotic species	Biological control is an innovative, cost-effective and eco-friendly approach.

Efficient use of natural resources

Action	Mitigation
Overexploitation/ habitat destruction	Using resources more efficiently, deriving maximum output per unit e.g. wood to pulp to paper.
Pollution	Develop sustainable strategies to promote the use of finite resources more efficiently.
Introduction of exotic species	Biological control.

Key points

Mitigation is possible by:

- Avoiding the impact altogether by not taking a certain action.
- Minimising impacts by limiting the degree or magnitude of the action.
- Rectifying the impact by repairing, rehabilitating or restoring the impacted environment.
- Reducing or eliminating the impact over time by preservation and maintenance.
- Compensating for the impact by replacing or providing substitute resources or environments.

 Exam tip

The ability to differentiate between the eco approaches to environmental conservation and preservation is key.

Did you know?

More than 2 million people across the United States have pledged to take part in the ENERGY STAR pledge; this would save US$18 billion in annual energy costs and prevent greenhouse gases equivalent to the emissions from 20 million cars.

Figure 2.23.1 *Traffic congestion in Port of Spain, the capital city of Trinidad and Tobago*

Causes of urbanisation

The process of urbanisation is often linked to increasing industrialisation. Rapid development of suburban areas and increased rates of rural to urban migration have contributed greatly to the increase in urban population, with a subsequent decline in rural numbers.

The main causes of urbanisation are:

1 Increase in population size due to natural increase caused by a decrease in death rates while birth rates remain high.

2 Rural neglect – in many developing countries development is focused in the city and town areas, and as a result very little financial input is available for the improvement of rural areas. In Jamaica the driving force of urbanisation is rural poverty, and the country's urban population now accounts for 54 per cent of its 2.7 million inhabitants.

3 Industrialisation and urban expansion act as a stimulus for the movement of people into an area. There is a widespread perception that urban areas are economically viable and therefore large numbers of poor people are drawn into these areas in the hope of achieving financial security. In Trinidad and Tobago there is a drive to develop areas outside of the capital to alleviate the pressures placed on available services.

4 Rural to urban migration and/or immigration is due to rural neglect, changes in lifestyle, job opportunities and the possibility of acquiring a higher standard of living. The current rate of migration into the city areas is high, as evident from the increased numbers of squatter settlements, slums and ghettoes that surround the cities.

Environmental impacts of urbanisation

Traffic congestion

Increased vehicular traffic in the Caribbean is due primarily to the volume of people working and moving through the cities, as well as the shortage of off-street parking, which means that people park indiscriminately on roadways. Public transport, while in demand, is often unreliable, less convenient or unavailable but the number one cause of traffic congestion is the drastic increase in the number of cars on the road annually.

The result is slow-moving or stationary traffic which exposes road users to severe air pollution from exhaust fumes. The increase in auto emissions of carbon monoxide and other hydrocarbons into the atmosphere has the twofold effect of reducing air quality and increasing the heat concentration, especially in urban areas.

Pollution

Environmental noise pollution is a threat to the health and well-being of urban inhabitants both in magnitude and severity because of population growth. Sustained rises in noise levels could lead to irreversible hearing loss, and has a detrimental effect on sleep, concentration, communication and recreational activities. Noise pollution is a growing concern in Trinidad and Tobago due to the increase in local vessel traffic, especially large commercial vessels with loud engines. The oil industry also contributes to ocean noise pollution, especially during oil exploration exercises, but there is a growing attempt to reduce the impact of these activities on turtles and cetaceans by suspending seismic activities in some cases when these animals approach the work area.

Health care

Rapid urbanisation has significant repercussions on the health status of city dwellers. City living and its increased pressures of mass marketing, availability of unhealthy food choices and easier access to automation and transport all have an effect on lifestyle, leading to an increase in non-communicable diseases (cardiovascular diseases, cancers, diabetes and chronic respiratory diseases).

Unplanned and rapid urbanisation can pose particular health risks, especially if the health care facilities in the urban areas concerned have not developed to keep pace with the increase in population. Many of the rural to urban migrants are poor, and their health status may not be good when they arrive. Living in close proximity, often in crowded and insanitary conditions, can give rise to diseases, and help spread communicable diseases.

Water supply

The demand for a clean, adequate and reliable supply of water often exceeds the ability of countries in the Caribbean to supply. Trinidad and Tobago boasts one of the most effective water and sewerage systems anywhere in the Caribbean, suggesting that the United Nations' Millennium Development Goal (set for 2015) with respect to achieving adequate drinking water and sanitation arrangements has already been achieved. An estimated 92 per cent of the 1.3 million inhabitants currently have access to safe drinking water. However, this supply is only available to around 26 per cent on an uninterrupted, 24-hour basis. The remainder of those supplied receive water on a weekly schedule, while a number of communities are still without a regular supply. Some 30 per cent of the population has a sewerage connection, while 58 per cent rely on soak-away or septic tanks.

Water pollution

In the Caribbean water pollution is widespread and caused mainly by human activities resulting from littering, unsettled quarry wash, industrial plant and factory waste, overuse of fertilisers and pesticides, bush fires, slash and burn and deforestation, and untreated or poorly treated sewerage.

Housing

The carrying capacity of many urban areas has been exceeded due to the increase in economic activities resulting from the establishment of large numbers of informal settlements around these urban centres. Across the Caribbean Region squatting is a real problem. In Jamaica former Housing Minister Dr Horace Chang indicated that there are approximately one million squatters in the country. The government is moving to enact legislation to make squatting a criminal offence. But even as the administration moves to adopt a tough policy on squatting, it cannot ignore the fact the large chunk of the population is in dire need of shelter. As most of those people are impoverished, unemployed and unskilled, it is difficult for them to find alternatives to squatting.

Figure 2.23.2 *Water pollution on our shores*

Did you know?

Jamaica: estimated 595 informal settlements; 75 per cent on State lands.

Trinidad: estimated 250 informal settlements on State land; 23,000 households applied for regularisation.

Guyana: estimated 215 sites, almost half of which deemed 'zero-tolerance sites'.

Informal settlement also a significant issue in Haiti, Grenada, Bahamas, St Vincent and Grenadines, Montserrat, St Lucia, St Kitts and Nevis, etc.

Key points

■ In the Caribbean the negative impact of urbanisation greatly outweighs the positive impact.

2.24 The impacts of urbanisation in the developed world

Figure 2.24.1 *Traffic congestion in Beijing*

Traffic congestion

Today, countries all over the world are faced with problems in their cities resulting from rapid urbanisation, such as traffic congestion, noise, water and air pollution, poverty, heavy demands on health care and housing resources. In China traffic jams are a way of life but on the Beijing–Tibet Expressways just outside of Beijing, drivers were trapped in a 62-mile traffic jam that lasted over 12 days in August 2010. Crawling along at a speed of just two miles per day, some drivers estimated that it took them three days to pass through the congestion. Traffic woes are a result of high traffic densities caused by too many vehicles on the road at the same time. The cost of traffic congestion in Japan alone annually is estimated to reach 12 trillion yen (US$131 billion, 1.07 billion Euro). Traffic jams not only induce stress, but also waste fuel and have an impact on the environment through increase gas consumption and increased car exhaust emissions.

Noise pollution

The main sources of environmental noise pollution are air conditioners and industrial equipment; noise from industry, construction, and demolition; noise generated by human activity such as lawn mowers or leaf blowers, loud music, barking dogs, children playing, and outdoor events such as concerts or festivals. Another significant source is transportation-related, with buses, trains, cars, motorcycles, trucks, and emergency vehicle sirens being the most significant causes of noise in urban areas.

In the United States, Canada, Europe, and most other developed parts of the world, different types of noise are managed by agencies responsible for the source of the noise. Bodies responsible for noise pollution reduction usually view noise as an annoyance rather than a serious problem, and reducing that noise often hurts the industry financially, therefore little is currently being done to reduce noise pollution in developed countries.

Health care

'The world is rapidly urbanising with significant changes in our living standards, lifestyles, social behaviour and health', says Dr Jacob Kumaresan, director of the World Health Organisation's Centre for Health Development based in Kobe, Japan. 'While urban living continues to offer many opportunities, including potential access to better health care, today's urban environments can concentrate health risks and introduce new hazards.'

Urban environments tend to discourage physical activity and promote unhealthy food consumption. Participation in physical activity is made difficult by a variety of urban factors including overcrowding, high-volume traffic, heavy use of motorised transportation, poor air quality and lack of safe public spaces and recreation/sports facilities.

Rapid urbanisation has significant repercussions on the health status of city dwellers. City living and its increased pressures of mass marketing, availability of unhealthy food choices and easier access to automation and transport all have an effect on lifestyle, leading to an increase in non-communicable diseases (cardiovascular diseases, cancers, diabetes and chronic respiratory diseases). Other risks faced by urban dwellers are obesity relating to unhealthy diets and the lack of physical inactivity, as well as the harmful use of substances and the risks associated with disease outbreaks. All these factors can impact negatively on the longevity of urban dwellers.

Water supply

Australia's rapid population growth and urbanisation are increasing the demand for water, energy and food, as well as the generation of waste streams, stormwater runoff, and the flow of nutrients and contaminants into the waterways. To keep providing high quality urban water services while maintaining water ecosystems and reducing the carbon footprint of humans, a more integrated urban water management system that would harness water, wastewater and stormwater sources is vital. Australia's large cities are currently investing more than A$30 billion in new water supplies, and most are diversifying their supplies away from rainfall-dependent storages to include desalinated water, decentralised supplies and some form of water recycling.

Water pollution

The European Commission has referred France to the European Court of Justice for the second time over persistently high levels of nitrate pollution found in drinking water in Brittany caused by runoff from intensive farming and bad agricultural practices. The Court stated that 37 rivers in Brittany contained concentrations exceeding this level. Since then, France has implemented a small number of measures aimed at reducing the volume of nitrates spread over agricultural land. However, those measures have proved to be insufficient for nine rivers. The Commission proposes asking the Court to impose on France a lump-sum fine of over €28 million and a daily penalty payment of €117,882. Many households in Brittany refuse to drink tap water because of fears of pollution levels, preferring to buy bottled water from supermarkets.

Housing

According to the United Nations Human Settlements Program (UN-Habitat), 'the worsening state of access to shelter and security of tenure results in severe overcrowding, homelessness, and environmental health problems'. The urban poor and large segments of low and moderate income groups have no choice but to rely on informal land and housing markets for access to land and shelter, thus fostering the expansion of irregular settlements in cities. Informal land and housing delivery systems remain the only realistic alternative for meeting the needs of low-income households.

The US has its own phenomenon of urban-homesteading where, in neighbourhoods with large numbers of uninhabited abandoned homes, poorer residents take over and refurbish properties for their own use. Usually a grass-roots endeavour, urban homesteading is legal and has been used by some state governments as a solution to the lack of affordable housing. There is also a growing squatter movement in the US following the financial crisis of 2007, which has led to thousands of homes being repossessed.

Did you know?

Japan has one of the fastest urbanisation rates in the world. Of all 47 provinces in Japan, only seven have populations under one million people.

World Health Day marks the anniversary of the establishment of the WHO in 1948. It is celebrated annually on 7 April with a different theme each year, featuring an area of priority in global public health.

✓ Exam tips

You should be able to discuss the environmental and social impact of urbanisation on developed countries.

Key points

- The impact of urbanisation is greater on areas that are already considered urban.

- Choice and decisions made by people affect the rate of urbanisation and where it occurs.

Module 2　Exam-style questions

Multiple-choice questions

1　Which of the following indices of poverty are based on longevity, education and income for women alone?
   A　Gross National Product.
   B　Gross Domestic Product.
   C　Human Development Index.
   D　Gender Development Index.

2　The difference between birth rate and death rate is called:
   A　the replacement fertility rate.
   B　doubling time.
   C　life span.
   D　natural increase.

3　When the death rate of a country equals the birth rate, the population size:
   A　decreases.
   B　increases slowly.
   C　remains the same.
   D　increases rapidly.

4　The LEAST successful method of controlling a country's population size is:
   A　birth control.
   B　financial incentives.
   C　natural disasters.
   D　government quotas on children produced.

Refer to the following table of population growth rates for Jamaica and Trinidad and Tobago in 2009 to answer items 5 and 6.

Country	Population (millions)	Annual Growth Rate (%)
Jamaica	2.7	1.2
Trinidad and Tobago	1.3	0.3

5　The estimated population of Jamaica for the year 2010 is:
   A　32,400
   B　320,400
   C　2,732,400
   D　2,667,600

6　The estimated doubling time for Trinidad and Tobago is:
   A　33 years
   B　23 years
   C　233 years
   D　30 years

7　A country is MOST likely to be sparsely populated in which of the following areas?
   I　Flat, lowland plains.
   II　Mainly commercial farming areas.
   III　Regions with intensive farming.
   IV　Regions with fertile, deep soils.

   A　I and II only.
   B　I and III only.
   C　I, II and IV only.
   D　II only.

8　Basic family-planning policy in most countries includes all of the following except:
   A　limiting families to two children each.
   B　providing information about prenatal care.
   C　helping parents space births as desired.
   D　helping parents regulate family size.

9　The age structure of a population is the number or percentage of:
   A　females age 14 years or under.
   B　females age 15 to 44.
   C　males age 15 to 44.
   D　persons of each sex at each age level.

10　Which of the following countries would produce the greatest rise in population size from experiencing a growth rate of 1.2%?
   A　country A, with a population of 100,000.
   B　country B, with a population of 1 million.
   C　country C, with a population of 10 million.
   D　country D, with a population of 1 billion.

Essay Questions

1　The following table shows the doubling time and per capita GNP for three countries:

Doubling Time and Per Capita GNP

Country	Doubling Time (years)	Per Capita GNP (US $)
Haiti	33	310
Jamaica	40	1600
St Kitts and Nevis	69	5870

a Define the terms **i** per capita GNP **ii** doubling
   time [2 marks]
b **i** Using the data presented in the table above,
   state the relationship between the rate
   of population growth and per capita GNP.
   [2 marks]

   ii Explain the relationship identified in (i) above.
   [4 marks]
c Briefly describe how the world population has
   grown from 1800 to the present. [3 marks]
d Outline THREE reasons for the growth in world
   population from 1800 to the present. [6 marks]
e Discuss ONE impact of rapid population growth
   on the environment. [3 marks]

   Total 20 marks

2 a Define the term 'sustainable development'.
   [1 mark]
b Discuss the impact of high fertility rates on
   a country's ability to achieve sustainable
   development. [6 marks]
c **i** Suggest an appropriate fertility rate that
   Caribbean countries should try to achieve.
   [1 mark]

   ii Justify your choice of fertility rate in i) above.
   [3 marks]
d Describe ONE strategy employed by EACH of:

   i a European country and **ii** a Caribbean
   country to reach an appropriate fertility rate.
   [6 marks]
e For ONE method identified in d) above, assess its
   effectiveness. [3 marks]

   Total 20 marks

3 Figure 2.25.1 illustrates the relationship between
poverty, population growth and environmental
degradation:

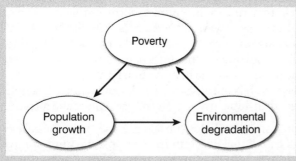

Figure 2.25.1 The cycle of poverty

a Discuss the relationship among the three
   components shown in Figure 2.25.1. [6 marks]
b State THREE social impacts of population growth.
   [3 marks]
c 'In order to reduce population growth in
   developing countries, the education of girls is
   especially important.'
   Evaluate this statement and in your evaluation,
   present at least FOUR reasons to support your
   position. [8 marks]
d Describe ONE impact of overconsumption on the
   environment in a developing country. [3 marks]

   Total 20 marks

4 The following table presents data for water
consumption per capita for developed and
developing countries.

Water consumption per capita

Group of countries	Water consumption per capita (m³)
Developed	500
Developing	993

a Calculate the percentage difference in the per
   capita water consumption between developed
   and developing countries. [4 marks]
b Describe ONE environmental and ONE social
   impact associated with high per capita water
   consumption. [4 marks]
c Account for the difference in per capita water
   consumption between developed and developing
   countries. [6 marks]
d **i** Define 'sustainable development'. [2 marks]

   ii Describe how achieving sustainable
   development can be hindered by the
   overconsumption of water resources.
   [4 marks]

   Total 20 marks

3.1 Natural resources

∞ Links

See 3.3–3.6 for the major categories, location and distribution of the natural resources in the Caribbean.

Natural resources are the essential elements of the natural environment, which are exploited to satisfy human wants. Like all parts of the environment, they have varying levels of biodiversity depending on the ecosystem, and they provide the basis of the livelihood of most of the populace, as well as the source of the GDP for many states. These resources are usually extracted in the form found in nature and then processed into desired products for human use and consumption. There are various methods of categorising natural resources, including source of origin and by their renewability. On the basis of origin, resources may be divided into those arising from abiotic sources, and those arising from biotic sources

Abiotic natural resources

Abiotic natural resources are those which come from nonliving, non-organic material. Examples of such natural resources in the Caribbean region include soil, land, beaches, fresh and marine water, air, minerals such as gold, heavy metals and ores such as bauxite.

Biotic natural resources

Biotic resources, on the other hand, originate from the living components and organic matter of the biosphere. These resources therefore include the ecosystems and species introduced in Module 1, and are crucial in the cycling of materials and nutrients in nature. Examples of biotic natural resources include forests (Figure 3.1.1), coral reefs, fish, birds and wildlife, and any materials that can be obtained from these primary resources. Because they are formed from decayed organic matter, fossil fuels such as natural gas and petroleum are also included in this category.

Extraction of natural resources

Natural resources are extremely important to the GDP of a country. Along with agriculture, which utilises soil or water, extractive industries are the basis of the primary sector of the economy in the Caribbean region. Resources are extracted or removed from nature in their primary form, so that they can be utilised by humans in a variety of ways. This often includes the processing of the resource into secondary resources: for example the processing of oil into plastics, paints, fertilisers and other by-products of the oil and gas industry. Extraction obtains the natural resource in its natural form, which is then processed to add value, which is the basis of secondary industries. Examples of extractive industries are fishing, mining, oil and gas drilling and forestry.

Extraction may be on a small scale for subsistence use, as in the case of most indigenous and traditional communities, or on a medium or large scale for commercial purposes. An example of commercial extraction is the mining of bauxite in Jamaica, limestone in Barbados, and oil and gas in Trinidad and Tobago. The process of extraction can utilise simple or complicated technologies, and may often have negative effects on the natural and human environment. As a result, the extraction of natural resources needs to be managed and tools for conservation are employed to ensure their sustainable use.

Management and conservation of natural resources

The extraction and exhaustion of natural resources is considered to be a sustainable development issue. In 1982 the United Nations published the World Charter for Nature, which underscored the need to protect nature from further depletion due to human activity. This was followed in 1985 by the Brundtland Commission's Report: *Our Common Future*, which stressed the importance of sustainable development.

Management and conservation of the natural resource depends on the type and origin of the resource. Biotic resources such as forests and coral reefs are generally managed by habitat conservation efforts, to conserve, protect and restore, habitat areas for wild plants and animals, especially conservation-reliant species, and prevent their extinction, fragmentation or reduction in range. Species such as fish are managed to ensure that they are harvested sustainably, and wildlife are protected using in-situ or ex-situ means to maintain their numbers and prevent their extinction.

Figure 3.1.1 *This forest in Belize is an example of one of the most diverse and important ecosystems in the Caribbean region*

∞ Links

See 3.19 for an explanation of in-situ and ex-situ conservation measures.

Key points

- Natural resources are substances that occur naturally in the environment, and are utilised by mankind in his everyday life.

- Natural resources may be categorised by their source of origin, and by their renewability.

- The origin of natural resources may be either biotic or abiotic, while a resource may be either renewable or non-renewable depending on its physical characteristics.

- The extraction of natural resources is the primary way in which mankind utilises these resources, and can have impacts on the human and social environment.

- As a result, various methods of management and conservation exist to address the issue of extraction and use of natural resources.

Learning outcomes

On completion of this section, you should be able to:

■ distinguish between renewable (inexhaustible) and non-renewable (exhaustible) natural resources

■ distinguish between consumptive and non-consumptive natural resources

■ understand the distinction between the two systems of categorising natural resources.

Did you know?

Tide, wave wind and air power are known as inexhaustible resources as they will never run out. Inexhaustible resources are also renewable resources, but not all renewable resources are inexhaustible. An exhaustible renewable resource needs time to regenerate when harvested.

Renewable and non-renewable natural resources

A natural resource may be categorised by its ability to regenerate itself when extracted. A natural resource may be either renewable or non-renewable.

Renewable natural resources

Renewable natural resources are those natural resources which have short cycle times, so can be replenished naturally. Some of these resources, like sunlight, waves, air, wind, etc., are continuously available and their quantity is not noticeably affected by human consumption. However, most renewable resources such as forests, fisheries, wildlife and coral reefs need time to regenerate when harvested. Therefore, if the renewable resources do not have such a rapid replenishment rate, or are harvested at rates which do not allow the resource to replenish, these resources become susceptible to depletion by over-use. Therefore, resources are renewable only so long as the rate of replenishment/recovery exceeds that of the rate of consumption.

Non-renewable natural resources

These are natural resources which were formed naturally in the environment, and occur in fixed quantities. Once they are utilised, they cannot regenerate, or are very slow to regenerate – so the resource does not replenish in the average lifetime of a human. Minerals and fossil fuels are the most common resources included in this category. Of these, the metallic minerals can be re-used by recycling their shape and form, but fossil fuels such as natural gas are exhausted when they are used.

Figure 3.2.1 *The Pitch Lake at La Brea in southwest Trinidad is the largest natural deposit of asphalt in the world: the lake covers about 40 ha, is reported to be 75 m deep and is a non-renewable resource*

Consumptive and non-consumptive use of natural resources

Natural resources may be either harvested and consumed, or be utilised in their natural state. These different methods of using natural resources are described as consumptive and non-consumptive.

Consumptive use of natural resources

The 'consumptive' use of a natural resource refers to extracting a natural resource from its source, and utilising it so that the amount used no longer exists. Each time that resource is used, its supply is reduced. Examples of consumptive use of natural resources in the Caribbean region include logging, quarrying and fishing. For example, when harvested by fishers, banga-mary (*Macrodon ancylodon*) or dolphinfish (*Coryphaena hippurus*) are taken from the sea, sold in a fish market, supermarket or restaurant to be eaten, and therefore no longer exist.

Non-consumptive use of natural resources

The non-consumptive use of a natural resource refers to the utilisation of a natural resource without reducing its supply. This is of course not possible for every natural resource. Examples of non-consumptive use include bioprospecting, ecotourism, research and watching wildlife: it is possible to enjoy the beauty of the wildlife such as whales without killing or removing them from the wild. The resource is being used for enjoyment by humans, and a benefit is derived from its use, even though it is not consumed. In the Caribbean, examples of the non-consumptive use of natural resources in this way include whale-watching off the coast of Scott's Head Dominica, ziplining in the forests of Dennery, St Lucia or walking across the canopy-walkway in Iwokrama, Guyana.

Figure 3.2.2 Red snappers (Lutjanus campechanus) for sale in Gouyave Market, Grenada: fishing is an example of the consumptive use of natural resources

Figure 3.2.3 Whale watching off Scott's Head on the coast of Dominica is an example of a non-consumptive use of a natural resource

Summary questions

1 Cite TWO examples of 'renewable' and 'non-renewable' natural resources.

2 With reference to one of the examples cited above, illustrate how it may be used in a 'consumptive' manner, and in a 'non-consumptive' manner.

3 Do you think all natural resources can be used both consumptively and non-consumptively? Explain.

Key points

- Natural resources may be categorised according to whether they are renewable (non-exhaustible) or non-renewable (exhaustible).

- They may also be characterised by their usage – if they are utilised to satisfy human needs, this is considered a consumptive use of the natural resource, but if humans can benefit from the resource without harvesting or extracting the resource then this is known as a non-consumptive use of the resource.

- Because these categorisations are based on different properties of the resource, they can be used simultaneously: for example, the red snappers in the picture may be considered a consumptive use of a renewable resource.

- The distinction between these two categorisations is an important learning point.

∞ Links

See later sections for factors affecting natural resource use in the Caribbean region.

3.3 Natural resources in the Caribbean region: Biodiversity I

Biodiversity

Biodiversity refers to the assortment of species and ecosystems on the planet, but the term can also be used to refer to smaller geographic areas. Thus, as mentioned in 3.1, the Caribbean is referred to as a biodiversity 'hotspot' because it has a high level of endemics, and has a high degree of primary vegetation. Biodiversity is a measure of the health of an ecosystems, and the general rule is that the greater the species diversity, the more stable the ecosystem. Tropical regions such as the Caribbean tend to have a higher level of biodiversity than other biomes such as temperate and polar areas, because they host highly diverse ecosystems. For example, two of the world's most diverse ecosystems: forests and coral reefs, are found in the Caribbean.

Species

The species diversity and genetic diversity of the Caribbean region constitutes one of its most important natural resources. Species may be flora like the greenheart (*Chlorocardium* spp.) found in the forests of Guyana or the West Indies mahogany (*Swietenia mahagoni*) – an endangered species native to Bahamas, Cuba, Jamaica, and Hispaniola. Examples of fauna include the beautiful reef-fish such as parrotfishes (Scaridae) in Bucco Reef off Tobago or grunts (Haemulidae) off the coast of Soufriere in St Lucia, and threatened species such as the marine turtles and the commercially valuable queen conch (*Strombus gigas*). Fish for human consumption such as the flying fish (order Beloniformes) or Atlantic blue marlin (*Makaira nigricans*) – a species of marlin endemic to the Atlantic Ocean – also form part of the Caribbean's vast array of species.

Ecosystems

The ecosystems of the Caribbean provide a number of important functions for the people of the Caribbean region. Ecosystems can be terrestrial, such as the forests in most of the Caribbean, or aquatic. Aquatic ecosystems may be divided into marine or saltwater ecosystems or freshwater ecosystems. Examples of ecosystems found in the marine environment include coral reefs and seagrass beds. There are also coastal ecosystems which include wetlands and mangroves. The Caribbean's ecosystems provide a wide range of ecosystem services, such the cycling of nutrients and elements, and are home to the Caribbean's many species. They therefore serve a crucial role.

Coral reefs are known as 'rainforests of the sea' as they can form some of the most diverse ecosystems on Earth. Because of their characteristics, these reefs are found exclusively in the warm, clear waters of the Caribbean Sea, as opposed to the colder waters of the western Atlantic coast of South America. Seagrasses are flowering plants which grow either partially or fully submerged in marine environments. They occur in shallow and sheltered coastal waters anchored to the bottoms, as they need to photosynthesise. They are therefore limited to growing in the submerged photic zone, and require clear water that is free of sediments. Seagrass beds are highly diverse and productive ecosystems, and are also a very important link in the food chain, for fish, and especially for herbivorous species such as the green turtle (*Chelonia mydas*) and manatees. Forests are also found on most Caribbean islands, although most have significantly reduced primary coverage, or secondary coverage. States with appreciable remaining cover include Guyana, Belize, Suriname and Jamaica, and this forest ecosystem has important functions for watershed management, climate regulation and the cycling of nutrients.

Freshwater ecosystems include rivers and streams. In the interface between the terrestrial and marine environment are coastal ecosystems such as mangroves and wetlands. A wetland is a land area that is saturated with water, either permanently or seasonally, and wetlands are important spawning grounds for fish and a filter for water and sediments. Mangroves, which are a distinctive wetland species of shrubs that grow in saline coastal sediment, are also another important coastal ecosystem. Mangrove swamps protect coastal areas from erosion and storm surge (especially during hurricanes), since their massive root systems are efficient at dissipating wave energy.

 Links

See 3.4 Natural resources.

Key points

- Biodiversity refers to the assortment of species in ecosystems and is a measure of the health of an ecosystem.

- The general rule is that the greater the species diversity, the more stable the ecosystem.

- The Caribbean region hosts many terrestrial and aquatic species.

- Ecosystems can be terrestrial, such as the forests in most of the Caribbean, or aquatic – aquatic ecosystems may be divided into marine or saltwater ecosystems and freshwater ecosystems.

- Examples of the main ecosystems found in the marine environment include mangroves, coral reefs and seagrass beds.

- The ecosystems of the Caribbean provide a number of important functions for the people of the Caribbean region.

Conservation status

The conservation status of species is important to understanding the numbers and distribution of a species in a geographic area, or entire planet. The IUCN *Red List of Threatened Species* is the best-known worldwide conservation status system. Species are classified by the IUCN *Red List* into nine groups according to criteria such as rate of decline, population size, area of geographic distribution, degree of population and distribution fragmentation. Species whose numbers are declining rapidly are categorised by the official term 'threatened', which is a grouping of three categories: 'vulnerable', 'endangered', and 'critically endangered'.

IUCN Red List categories, explanations and examples

Category and explanation	Examples
Critically endangered (CR) Extremely high risk of extinction in the wild.	The Grenada dove (*Leptotila wellsi*) is endemic to Grenada and the national bird of Grenada; it is considered to be one of the most critically endangered doves in the world (Bird Life International, 2000). The Jamaican iguana (*Cyclura collei*) was thought to be extinct until a small population of about 200 individuals was rediscovered in 1990 in the Hellshire Hills of Jamaica. The 'mountain chicken' or the giant ditch frog (*Leptodactylus fallax*) is eaten in Dominica and Montserrat: and has been rapidly declining in numbers also from habitat loss and disease.
Endangered (EN) High risk of extinction in the wild.	The West Indian mahogany (*Swietenia mahagoni*) is the species from which the original mahogany wood was produced; it is native to the Bahamas, Cuba, Jamaica and Hispaniola. The Nassau grouper (*Epinephelus striatus*) is fished both commercially and for sport; it is the most important of the groupers for commercial fishery in the West Indies, but has been endangered by overfishing. The imperial parrot (*Amazona imperialis)* or sisserou is the largest member of the diverse genus *Amazona*: it is endemic to mountain forests in Dominica, and is the national bird, featured on the Dominican flag. One of the last remaining habitats of the parrot is on the slopes of Morne Diablotins, which was designated a national park in 2000.

Vulnerable (VU) High risk of endangerment in the wild.	The West Indian manatee (*Trichechus manatus*), whose range is throughout the Caribbean, is the largest surviving member of the aquatic mammal order Sirenia, which also includes the dugong and the extinct Steller's sea-cow. It is divided into two sub-species: the Florida manatee (*T. m. latirostris*) and the Antillean or Caribbean manatee (*T. m. manatus*). The bigleaf mahogany (*Swietenia macrophylla*), found in Belize and Central America, is one of three *Swietenias* which are the American or true mahoganies. The St Vincent parrot (*Amazona guildingii*), which is endemic to St Vincent and the Grenadines, has declined because of habitat loss and because it is hunted for food and trapping for the cage-bird trade. The Nicholas Wildlife Aviary Complex, located within the Botanic Gardens of St Vincent, maintains a vital captive breeding and conservation programme to conserve the St Vincent Parrot, and it is listed on Appendix I and II of *CITES* (i.e. it is split-listed). The St Lucia parrot (*Amazona versicolor*) is endemic to St Lucia and is the country's national bird. It is threatened by habitat loss: the species declined from around 1000 birds in the 1950s to 150 birds in the late 1970s, at which point a conservation programme began to save the species. By 1990 the species had increased to 350 birds.
Near threatened (NT) Likely to become endangered in the near future	The Caribbean reef shark (*Carcharhinus perezi*), found throughout the Belize Barrier Reef and the tropical western Atlantic and Caribbean, is one of the largest apex predators in the reef ecosystem, and is the most commonly encountered reef shark in the Caribbean Sea. Its population has declined off Belize and Cuba from overfishing, and exploitation continues in other regions. Because of their significance to ecotourism (a profitable ecotourism industry has arisen around this species involving organised 'shark feeds'), they are protected in the Bahamas by national law. The Caribbean coot (*Fulica caribaea*) is a large waterbird which is a resident breeder in the Caribbean and parts of Venezuela. Populations have suffered a marked decline throughout the Caribbean as a result of hunting pressures (including the taking of eggs for local consumption), pollution, habitat degradation and introduced predators.

Key points

- The conservation status of species is important to understanding the numbers and distribution of a species in a geographic area.
- The IUCN *Red List of Threatened Species* is the best-known worldwide conservation status listing and ranking system.
- Species are classified into nine groups.

Red List categories

There are nine of these as follows:

- Extinct (EX) – No known individuals remaining.
- Extinct in the Wild (EW) – Known only to survive in captivity, or as a naturalised population outside its historic range.
- Critically Endangered (CR) – Extremely high risk of extinction in the wild.
- Endangered (EN) – High risk of extinction in the wild.
- Vulnerable (VU) – High risk of endangerment in the wild.
- Near Threatened (NT) – Likely to become endangered in the near future.
- Least Concern (LC) – Lowest risk. Does not qualify for a more at risk category. Widespread and abundant taxa are included in this category.
- Data Deficient (DD) – Not enough data to make an assessment of its risk of extinction.
- Not Evaluated (NE) – Has not yet been evaluated against the criteria.

Figure 3.4.1 *Summary of 2006 IUCN Red List categories*

∞ Links

See 3.14 for more on species depletion and extinction.

3.5 Natural resources in the Caribbean region: Water

Did you know?

Grand Etang is a submarine volcano that is extinct, and evidence suggests that it is connected to Kick em Jenny – an active submarine volcano located off the coast of Grenada, because when Kick Em Jenny is bubbling, so too are the waters in Grand Etang Lake!

The importance of water

Water is an extremely valuable resource in the Caribbean, since it fuels agriculture, industry, tourism and domestic use. It provides various ecosystem services as well as providing water for man's many requirements. Additionally, water itself may provide a habitat for commercial species such as the Nassau grouper (*Epinephelus striatus*) or queen conch (*Strombus gigas*), both marine species, or tilapia (*Tilapia* spp.), which are freshwater species of fish widely used in the Caribbean. In addition, water is home to other biodiversity and wildlife, such as the marine turtles of the Caribbean Sea, marine mammals such as manatees, and cetaceans (whales and dolphins). Freshwater wildlife such as the Arapaima (Arapaima *gigas*) – the world's largest freshwater fish – and the endangered giant otter (*Pteronura brasiliensis*) may be found in rivers in Guyana.

The aquatic resources of the Caribbean may be classified into two broad headings: fresh water or marine water. The interface between marine and freshwater ecosystems is termed brackish water – water that has more salinity than fresh water, but not as much as seawater. It may result from the mixing of seawater with fresh water, as in estuaries and coastal wetlands.

Water salinity based on dissolved salts			
Fresh water	Brackish water	Saline water	Brine
< 0.05%	0.05% – 3%	3% – 5%	> 5%

Marine water is found in the sea and oceans, such as the Caribbean Sea and the Atlantic Ocean. It has a higher salt content than fresh water. Many of the trademark ecosystems in the region are associated with its marine resources, including coral reefs, seagrass beds and mangroves. Mangroves and wetlands are also to be found in brackish water, and constitute an important coastal ecotone in the region. Ecotones are transitional zones between two biomes, and they have important physical, ecological and economic functions.

Fresh water, which is divided into surface water and groundwater, is an important part of the water cycle (Module 1), and is found in rivers, streams and waterfalls. Surface water features include rivers, lakes and shallow pools such as the Emerald Pool in Dominica. Rivers form an integral part of the lives of the populace of the Caribbean region – they are sources of water for domestic, industrial and agricultural purposes; food in the form of fish and shellfish; a transport route; a place to do laundry and in some cases a social, cultural and religious meeting place. Rivers in Caribbean regions may be large – as those in Belize, Trinidad and Tobago, Suriname, Guyana and Jamaica, or small – for example the smaller islands of St Vincent and the Grenadines, St Lucia, Dominica and Grenada. Antigua & Barbuda and Anguilla, on the other hand, have very little surface water, which means they have to rely on the desalinisation of seawater, or the importation of water to acquire their fresh water.

Lakes are another form of aquatic resource in the Caribbean region, and may be formed in depressions, basins or when a river slows on its course,

Did you know?

Desalinisation is the process of removing salt and other minerals from saline water.

or in the case of many of the volcanic islands on the Antillean chain, in the craters of extinct volcanoes. Lakes are often relatively still bodies of water, which can be contrasted with rivers or streams, which are usually flowing. They may be fed either by rainfall, an underground aquifer or a river, and serve many purposes including as tourist attractions and a source of fresh water for the populace – an example of both is Grand Etang in Grenada. Dominica's Boiling Lake is situated in the Morne Trois Pitons National Park World Heritage site, and is the second largest hot lake in the world. The Great Salt Pond, located in St Kitts and Nevis, is another lake that provides a habitat for birds and fish, and is a nursery for other sea-dwelling fish.

Waterfalls in the Caribbean region include the Diamond Falls in St Lucia, Annandale Falls in Grenada and the Kaieteur Falls in Guyana. Many waterfalls are used for energy in the form of hydropower: for example the Government of Jamaica owns nine hydropower sites, and the Government of Guyana has proposed a plant at Amaila Falls. Most Caribbean islands, however, do not have water in sufficient amounts to utilise this form of energy.

Groundwater is another important aquatic resource, and may take the form of rivers and streams which are found below the surface of the soil. It is also an important source of potable water in Barbados. Other islands use surface water such as rivers and lakes for this potable water. Harrison's Cave, which is an important tourist site in Barbados, is an underwater cavern with stalactites, stalagmites and flowing groundwater.

Figure 3.5.1 *Wetlands are an important ecotone, and an interface between the marine and freshwater aquatic environments in the Caribbean region*

Key points

- Water is a very important natural resource for the Caribbean region, and is required by the people of the Caribbean countries in all aspects of their everyday life.

- Water is used for domestic, commercial, agricultural and industrial purposes, as a source of potable water and for cultural, spiritual and religious purposes.

- The region's aquatic resources may be divided into marine water, fresh water and brackish water.

- The marine environment is host to important ecosystems such as coral reefs and seagrass beds, which often give way to mangroves and wetlands which may be located in brackish waters.

- Brackish ecosystems are ecotones – and those that contain wetlands serve important functions for the Caribbean region.

- Fresh water may be in two main forms – surface waters such as rivers and lakes, and groundwater.

- Because water resources are so important, various methods of management and conservation exist to address the issue of the management and use of water resources.

Did you know?

The Mother Fall of the Trafalgar Waterfalls (which comprises the Mother Fall, the Father Fall and the Emerald Pool) in Morne Trois Pitons National Park, Dominica, is an example of one of the many majestic waterfalls in the Caribbean region. Morne Trois Pitons National Park has been a World Heritage Site since 1997.

∞ Links

See 3.16 for the impacts and threats to aquatic resources in the Caribbean region.

Learning outcomes

On completion of this section, you should be able to:

- identify the major categories of minerals and hydrocarbons in the Caribbean region
- identify the location of these resources in the Caribbean region.

Did you know?

Bauxite was named after the town of Les Baux, France, where it was discovered in 1821. Its colour ranges from dark red or brown to pink or nearly white, depending largely on the amount of iron oxide in it. Unlike copper, tin and lead, which have been used since ancient times, aluminium is a 'young' metal which has only been in use for about 150 years

Minerals and hydrocarbons

Hydrocarbons are a primary energy source for current civilisations, and their predominant use is as a combustible fuel source. A mineral is a naturally occurring substance that is solid and stable at room temperature, which usually occur in rocks or ores within the Earth's surface. Commercially valuable minerals and rocks such as bauxite, gold and limestone are referred to as industrial minerals, while gems are minerals with an ornamental value, and are distinguished from non-gems by their beauty, durability and, usually, rarity.

Hydrocarbons

Hydrocarbons are made of decomposed organic matter, consisting entirely of hydrogen and carbon, which were pressurised thousands of years ago into organic carbon and hydrogen compounds such as crude oil, natural gas and pitch. The majority of hydrocarbons found on Earth occur naturally in crude oil. They are economically important, because major fossil fuels such as petroleum and natural gas, and their derivatives such as plastics, paraffin, waxes and solvents, are heavily utilised in everyday life in the Caribbean region. However, because hydrocarbons release carbon dioxide into the atmosphere when they combust, they are one of the major sources implicated as contributing to global climate change.

In the Caribbean region, the Republic of Trinidad and Tobago is the only state that currently exploits hydrocarbons in any appreciable quantity. Other states such as Guyana and Suriname are currently exploring this option, while Jamaica and Barbados have exploited modest quantities. Most states are, however, heavily dependent on hydrocarbons and their derivatives, so they are an important natural resource to the Caribbean region. The 2005 Petrocaribe Agreement is an oil alliance of 18 Caribbean states (12 CARICOM states, excluding Barbados, Montserrat and Trinidad & Tobago plus six others) with Venezuela to purchase oil on conditions of preferential payment, and is an importance source of much fossil fuel utilised by Caribbean states.

Minerals

Minerals such as those of industrial importance, gems and precious metals are also found in the Caribbean region. For example, gold, diamonds and other precious and semi-precious gemstones are found in Guyana, gravel in Guyana, limestone in Barbados and Jamaica and bauxite in Jamaica and Guyana.

Gold is a bright, yellow-coloured precious metal, which is valuable, and highly sought after for coinage, jewellery, and other arts. Gold has a wide variety of uses, and because of the high demand for the metal, it is exported. Diamonds are also mined to be used in jewellery, industrial drillbits and other uses. Gold and diamonds are both found in abundance in Guyana and Suriname.

Bauxite, which is the name of any ore that has a large amount of aluminium hydroxide, is found in Jamaica, Guyana and Suriname.

Bauxite is converted to alumina, and then to aluminium metal: it takes 2–3 tonnes of bauxite to produce 1 tonne of alumina, and 2 tonnes of alumina to produce 1 tonne of aluminium metal. It is therefore a resource-intensive process, which requires a lot of energy. Aluminium has a wide variety of uses, including in the construction, transport, electrical and packaging industries, ranking aluminium as one of the top three used metals – only iron and steel (which is an alloy) are used more. The recycling of aluminium (see 3.23) is being promoted, since it is 100 per cent recyclable and requires 95 per cent less energy than the process required to convert bauxite to aluminium metal.

Figure 3.6.1 *Bauxite mining*

Sand and gravel, which are quarried in countries such as Dominica and Guyana, are important to the construction industry. However, the illegal mining of sand from beaches is a serious conservation issue in the region.

Key points

- Hydrocarbons and minerals are important natural resources for the Caribbean region, and their exploitation and use contributes a significant part of the GDP of the region.

- Hydrocarbons are the principal source of energy in the region, and many Caribbean countries import their supplies under agreements such as the 2005 Petrocaribe Agreement.

- Hydrocarbons also have a wide variety of derivative uses.

- Minerals may be categorised as those of industrial importance, such as bauxite, limestone and gravel; and precious metals and gems.

- They are generally extracted and exported in their primary state from the region.

- Because hydrocarbons and minerals are so important, various methods of management and conservation exist to address the issue of the management and use of these resources.

Did you know?

Although the famous geologist Sir Thomas de la Beche noted the presence in Jamaica of the red marly soil associated with bauxite and other minerals in 1827, and observations were made by another geologist, C. Barrington Browne, it was not until 1952 that bauxite mining commenced in Jamaica.

∞ *Links*

See 3.18 for the effects of exploiting minerals and hydrocarbons in the Caribbean region.

Soil

Soil is a precious resource in the Caribbean, as it forms the basis of the agriculture industry. The Caribbean states have historically relied on extensive agriculture – primarily of sugar cane, but also other crops such as arrowroot and cotton. Historically, the agricultural industry has been the mainstay of the region's GDP – beginning with the advent of 'King' sugar in the 1400s. Although the importance of sugar cane cultivation has declined, many states such as Guyana and Barbados still cultivate the crop for export, while others such as St Kitts and Nevis and St Vincent and the Grenadines have stopped production for export, but still grow to supply the raw material of the rum and spirit industry. Many countries also incorporate tours of old sugar plantations into their tourism package. An example is St Nicholas Abbey, a plantation house in Barbados, and one of only three genuine Jacobean mansions in the western hemisphere. Sugar has been grown on the plantation since 1640 and was processed on the property until 1947. There is still the evidence of the mill and sugar making edifices. Today, some states like Guyana and Suriname also cultivate crops such as rice for export, while St Lucia and the French West Indian islands of Guadeloupe and Martinique produce bananas. Jamaica is famous for its renowned Blue Mountain coffee, which is grown in the Blue Mountain range. The smaller islands on the Antillean chain of the Caribbean also produce ground provisions (roots and tubers), and Grenada is famous for its production of nutmeg.

Landscapes

Formations such as beaches, and land formations such as mountains, are important to the aesthetic beauty of the Caribbean region. Mountains are an important landform in the Caribbean because they serve many purposes. They are a watershed area, and many states derive their usable water from mountain areas. In addition, as in the case of Jamaica, they are used for agriculture, and also have tremendous aesthetic value. Cliffs, which are significant vertical or near-vertical rock exposures formed as a result of the processes of erosion and weathering, are common in the more mountainous Caribbean states. Cliffs may be found on coasts, such as those in Point Michel, Dominica, Anguillita Island, Anguilla or Negril Jamaica – where cliff jumping is a popular pastime; or in mountainous areas of the more mountainous islands such as Tobago, Grenada, St Lucia and St Vincent and the Grenadines.

Beaches are a trademark attraction of the Caribbean region. Examples of white sand beaches include Grand Anse in Grenada, Macaroni Bay in St Vincent and the beaches on Ambergris Caye in Belize. Dominica has black sand beaches, while Guyana has a series of nine beaches made up of shells. These beaches feature centrally in the region's image as a tourism destination, and have other important functions – including structural, economic and ecological. For example, many beaches in the Caribbean, such as Shell Beach in Guyana, Levera in Grenada and Matura in Trinidad, are important nesting sites for the six species of marine turtles common to the Caribbean region.

Figure 3.7.1 *The scenic Parlatuvier Bay, Tobago*

Seascapes

The Caribbean Sea and the Atlantic Ocean encircle the islands of the Caribbean, and lap the shores of the mainland countries of Guyana, Suriname and Belize. The marine environment comprises a large part of the territory of Caribbean states, and includes unique ecosystems and a wide range of biodiversity. It is important for fisheries, shipping and activities such as swimming, snorkelling and diving.

With the exception of Guyana, most of the Caribbean is washed by sparkling blue waters, which along with its beaches and its sunny climate, makes the seascape one of the region's most distinctive features. This sun-sand-and-sea combination is marketed as part of many tourism packages, since these inherent characteristics make the region a year-round destination for tourists all over the world. Notable features of the sea include picturesque bays such as Parlatuvier and Englishman Bays in Tobago, Jolly Harbour in Antigua as well as the dramatic landscape-seascape combination of archipelagos such as the Grenadines, the Bahamas, and indeed the entire Caribbean chain.

Key points

- Soil, landscapes and seascapes are important natural resources for the Caribbean region.

- These areas are an important aspect of the aesthetic appeal of the Caribbean region, and are therefore central to the region's tourism industry.

- Additionally, these areas host important ecosystems such as forests, mangroves and wetlands.

- As a result of their importance, various methods of management and conservation exist to address the issue of the management and use of soil, landscapes and seascapes in the region.

Natural resources and livelihoods in the Caribbean

Natural resources have been utilised as the base of livelihood and income for the people of the Caribbean region for centuries. The natural resources of the region are also crucial to the GDP and earning of foreign exchange for the Caribbean countries.

Tourism

The World Tourism Organisation (2005) defines tourism as 'travel to and staying in places outside one's usual environment for not more than one consecutive year for leisure, business and other purposes'. Tourism is the leading income earner for the Caribbean, with some studies identifying the region as being the most tourism-dependent in the world. Tourism in the Caribbean may be for recreation, leisure, or business purposes. Tourism brings in large amounts of income in payment for goods and services. The industry also creates derivative opportunities for employment in the service sector of the economy, including transportation services, such as airlines, cruise ships and taxis; hospitality services including hotels and resorts; and other facilities such as entertainment venues, amusement parks, casinos and shopping malls.

The tourism industry in the Caribbean capitalises on the inherent sun-sand-sea characteristics of the region, and employs thousands of people either directly or indirectly. This includes the hotel industry, restaurants, sightseeing, cottage industries and entertainment. It is therefore an important part of the economy of the region. In recent years the Caribbean has moved away from the mass-tourism model to niche and green tourism.

Ecotourism deals with living parts of the natural environments, and typically involves travel to destinations where flora, fauna and cultural heritage are the primary attractions. The tourism industries in Guyana, Dominica and Suriname are more geared towards ecotourism than some other countries in the region. Additionally, natural resource management and conservation (3.20) can be utilised as a specialised tool for the development of ecotourism. Ecotourism programmes can be introduced for the conservation of resources where there are threats from human encroachment on habitat and species depletion. Ecotourism may also play an important role in environmental education and awareness (3.29), since sustainable tourism must address the issue of ecotourists removed from the effects of their actions on the environment. Initiatives are often incorporated into the experience to improve the tourists' awareness, sensitise them to environmental issues, and foster positive attitudes about the places they visit and the species they may see. An example is a programme in the Folkstone Marine Park in Barbados, which has helped mitigate negative environmental impacts by providing information and regulating tourists on the parks' beaches used by nesting endangered sea-turtles.

Cruise ship tourism is another main form of tourism for the region, with many cruise ships using the states in the Caribbean as ports of call (Figure 3.8.1). However, because cruise ships offer amenities and activities on board, tourists balance their time between the onboard experience and visiting the attractions at various ports of call. Cruise ships have been identified as a key source of marine pollution and most states have signed the MARPOL Convention in an attempt to address this issue (see 3.30).

Fisheries

Fisheries is possibly the leading source of livelihood for the majority of the rural communities in the Caribbean region. Villages such as Six Men's Bay and Oistins in Barbados, Gouyave in Grenada and Scott's Head in Dominica are noteworthy 'fishing' villages. Fish is harvested in both deep and shallow areas of the sea in areas such as the Pedro Banks off Jamaica, and off the Atlantic shelf of South America. Fish is also harvested in rivers and streams in some states. Fish is a popular source of protein in the Caribbean, and is also exported, and in recent times the rate and extent of fishing has become a cause for concern. Many projects and strategies such as the maximum sustainable yield (MSY) have been employed to keep the resource sustainable.

Figure 3.8.1 A cruise ship in a Caribbean port

Other

Other forms of livelihoods in the Caribbean range from primary industries such as farming, to secondary industries such as the preparation of rum, Caribbean specialties, souvenirs and other activities. Even tertiary activities such as banking, financial and commercial services benefit from the exploitation of the many natural resources of the region.

Foreign exchange earner

As illustrated by the livelihoods which are derived from natural resources of the Caribbean region, they are a significant foreign exchange earner. In addition to foreign exchange earned in tourism, fish is exported and processed. Precious metals such as bauxite and gold constitute significant foreign exchange earners. Other sources of income include processed materials – all of which are sold in stores and airports or exported to foreign markets.

Key points

- The Caribbean region has many natural resources which play an important part in the livelihood and income of the populace of the region.
- Tourism in various forms: sun-sand-and-sea, ecotourism and cruise-ship, is one of the main sources of livelihood and income for the Caribbean.
- Fisheries is another important source of livelihood and income for the Caribbean, with many rural towns in the region being known as 'fishing towns'.
- Other sources of livelihood and income for the Caribbean region include secondary industries such as the preparation of rum, Caribbean specialties, souvenirs and other activities and tertiary activities such as banking, financial and commercial services.

The importance of natural resources: Food security and raw materials

On completion of this section, you should be able to:

- identify the role of natural resources in providing food and raw materials for the Caribbean region

- assess the importance of natural resources to food security and industry in the Caribbean region.

Did you know?

Mahogany is the national tree of the Dominican Republic and Belize, and appears on the national seal of Belize.

Livestock is an important part of the agricultural industry, and many Caribbean states rear poultry and other livestock for domestic consumption and the tourism industry. Chicken and eggs are reared in Barbados, Guyana and many territories to supplement imported food.

Some Caribbean countries have begun ventures into aquaculture and mariculture to rear commercially valuable species. Examples include the cultivation of tilapia (cichlids from three distinct genera: *Oreochromis*, *Sarotherodon* and *Tilapia*) in Guyana and Trinidad & Tobago, shrimp (*Penaeus vannamei*) in Belize, queen conch (*Strombus gigas*) in the Turks & Caicos Islands and the spiny lobster (*Panulirus argus*) in the British Virgin Islands.

Natural resources and food security in the Caribbean

Since the times of the indigenous peoples, the natural resources of the Caribbean have provided food, shelter and other raw materials for the populace. With the colonisation of the region, these resources now provide a plethora of services for the Caribbean countries, and play an important role in their food security. They also provide raw materials and service for the people of the region.

Agriculture

Agriculture has historically been the mainstay of the livelihood and economies of the Caribbean region. Apart from exporting various crops or secondary products derived from them, agriculture is a fundamental part of the food security of the Caribbean countries. Much of the produce grown in the region is consumed by the people there, and the ability of the region to grow its own food means that states can reduce the amount of their GDP they spend on importing food. Products such as rice from Guyana and provisions from the Organisation of Eastern Caribbean States (OECS) countries are examples of foods which are utilised in the region, or exported.

Fisheries

Fish and seafood, along with chicken, are the staple proteins in the Caribbean. In addition to fish, the region relies on its marine environment for a variety of products including lobster, conch and shrimps. Sea-turtles, sea urchins and seaweeds are often harvested – some in an illegal manner. Fish, shellfish and crustaceans are also harvested from freshwater sources in states such as Guyana and Trinidad and Tobago.

Aquaculture and mariculture

As a result of the demand for fish and seafood, and other factors such as pollution, habitat destruction and fragmentation, fisheries have been declining in the region. Some states are exploring the option of rearing commercially valuable species by means of aquaculture, mariculture or aquaponics.

Aquaculture (or aquafarming), is the farming of aquatic organisms such as fish, crustaceans, molluscs and aquatic plants. Aquaculture involves cultivating freshwater and saltwater populations under controlled conditions, and can be contrasted with commercial fishing, which is the harvesting of wild fish. Finfish and shellfish fisheries can be conceptualised as akin to hunting and gathering while aquaculture is akin to agriculture. Particular kinds of aquaculture include fish farming, shrimp farming, algaculture such as seaweed farming, and the cultivation of ornamental fish. Particular methods include aquaponics (or pisciponics), and integrated multi-trophic aquaculture or IMTA.

Mariculture is a specialised branch of aquaculture involving the cultivation of marine organisms for food and other products in the open ocean, an enclosed section of the ocean, or in tanks filled with seawater. Products of mariculture include marine fish, shellfish and seaweed, as well as non-food products such as fish meal and cosmetics. Mariculture has many social and environmental impacts.

Natural resources provide raw materials for industrial processes

Oil and natural gas

Oil and natural gas is an important resource in all aspects of life in the Caribbean. It is required for powering houses, fuelling cars, boats and aeroplanes, as well as industry. The region has a reliance on petroleum-based fuel, and most of the GDP of states is spent on it. This is because, with the exception of Trinidad and Tobago, no Caribbean state is exploiting petroleum in commercially significant quantities. Petroleum is refined and separated into a large number of consumer products, from gasoline to asphalt, and chemical reagents are added to make plastics and pharmaceuticals.

Forests

The products from forests are an important resource, both for the local industry and export. Forests provide many raw materials including timber, pharmaceuticals and non-timber forest products (NTFPs). Guyana, Belize and Suriname practise large-scale commercial logging for export, while most states use the products of their forests for domestic and fuelwood consumption. Non-timber forest products are also raw materials obtained from the forest. These products are highly promoted in sustainable forestry, because they are non-consumptive, as often they do not necessitate the harvesting of trees. Examples include wildlife, nuts, oils (e.g. from crabwood (*Carapa guianensis*) and medicinal plants. Forest resources in the Caribbean region are generally managed by legislation, because if they are not harvested sustainably (i.e. they are harvested in excess of the maximum sustainable yield), the resources can decline. An example is the drastic reduction of mahogany (*Swietenia* spp.) in Belize, Haiti, the Bahamas and Jamaica. Today, all species of *Swietenia* grown in their native locations are listed by CITES and are protected, with most commercially consumed species grown in forest plantations.

Other products

Other raw materials such as sand from Guyana and gravel from quarries in Guyana and Dominica are used in the construction industry. Sand is mixed with cement and sometimes lime to be used in masonry construction, and is often a principal component in concrete, a critical construction material in the region. For reasons of cost and availability, use of timber as a construction material has declined in the Caribbean.

Sandy soils are ideal for crops such as watermelons, and their excellent drainage characteristics make them suitable for intensive dairy farming. This must be managed carefully, since in some areas such as the Soesdyke region in Guyana, sandy areas are important catchment areas for freshwater supplies. Runoff from agriculture and livestock can pollute these areas, rendering the freshwater resources unusable. Sand is also graded, and used in sandblasting as an abrasive in cleaning, in sandbags to protect against floods, and in the filtration of water. Sand's many uses have given rise to a significant dredging industry, raising environmental concerns over fish depletion and flooding.

Figure 3.9.1 *Logs in their unprocessed form*

∞ Links

See 3.2 for an explanation of non-consumptive resource use.

Did you know?

Gravel is composed of unconsolidated rock fragments, and is an important commercial product, with a number of applications. These include construction uses and shoreline and river protection in the form of Gabion baskets. These are a cage, cylinder, or box filled with rocks, concrete, or sometimes sand and soil for use in civil engineering and road building across the Caribbean.

∞ Links

See 3.31 for more on CITES.

Key points

- The Caribbean region has many natural resources which provide food and raw materials.

- Agriculture is important for the food security of the population, as well as to provide for the tourism industry.

- Fish and seafood are crucial to the Caribbean's food security.

- Oil and natural gas and forests are important raw materials for industrial processes in the region.

Learning outcomes

On completion of this section, you should be able to:

- identify the role of natural resources in providing recreation for the Caribbean region
- assess the sacred and intrinsic value of natural resources in the Caribbean region.

Figure 3.10.1 *Underwater sculptures off Molinère/Beausejour in Grenada*

Recreational and aesthetic value of natural resources

Besides their many other uses, the Caribbean's vast array of natural resources and features also provide the backdrop for recreation by both Caribbean nationals and tourists. The combination of the natural resources and their aesthetic value foster activities, including sea-bathing in the waters of the Caribbean Sea, on beaches such as Maracas Bay in Trinidad, which is known for its bake and shark, to the beaches of Montego Bay in Jamaica. Some of the beaches on Jamaica's north coast were used in the filming of James Bond movies such as *Dr No*, and the filming of *Pirates of the Caribbean* took place in St Vincent and the Grenadines, Dominica and the Bahamas. The sea is also used for diving and snorkelling, and distinctive features such the Grenada Underwater Sculpture Park off Molinère / Beausejour are visited by many tourists each year. The rougher waters of the Atlantic coasts are used for surfing and parasailing, and some islands like Dominica and Antigua have whale-watching activities. Visitors can spelunk at Harrison's Cave in Barbados, hike through Dominica's Valley of Desolation or climb St Lucia's famous peaks, the Pitons. The rainforest also provides entertainment in the form of ziplining or canopy walkway, as found in St Lucia and Guyana respectively.

Figure 3.10.2 *Zip-lining in Dennery, St. Lucia*

Sacred and spiritual value

Many religions regard the environment as a sacred and spiritual resource. For example, the sea and rivers are used by the Hindus in Guyana and Trinidad and Tobago for prayer, planting jhandi (ceremonial) flags, and for funerary rites. Some Christian denominations also use these resources for baptism and other symbolic acts.

In territories where indigenous peoples still remain, the connection between their traditional way of life and the environment is interlinked. This is reinforced by legends such as that of Kai in Guyana, recent developments in Belize in the Aurelio Cal case and from anthropological studies of the life and customs of indigenous cultures.

Intrinsic value

Intrinsic value is the value that an entity possesses as a result of being itself, as opposed to its instrumental value, which is the value that something has on account of the demand for it, or its value. Ecosystems and species possess a wide variety of instrumental values (for example, cultural value, recreational value, medicinal value, spiritual value, transformational value, natural resource value, and ecosystem services value), and there is a school of thought that views ecosystems and species as also having non-instrumental value or intrinsic value. People value species and ecosystems intrinsically – for their complexity, diversity, spiritual significance, wildness, beauty, or wondrousness.

The idea that nature and biotic diversity have intrinsic value has been defended by several influential environmental ethicists (Soulé, 1985; Rolston, 1986; Callicott, 1989), and it has featured prominently in some significant international declarations regarding the environment (for example, Earth Charter International 2000). Those who endorse the view that species and ecosystems possess intrinsic value believe that recognition of this value is crucial both in justifying conservation biology and setting appropriate conservation goals. They also believe intrinsic value is relevant to developing particular conservation and management plans, strategies, and methods, since these need to reflect the values at stake. However, not all environmental ethicists agree that intrinsic value is crucial to justifying conservation goals and developing management plans and methods, and many rely on the instrumental value to guide these efforts.

Key points

- Natural resources play many roles for the populace of the Caribbean region.

- These include aesthetic and recreational roles which are not only important for the people of the Caribbean themselves, but for tourists who visit the region, and the GDP of the Caribbean countries.

- Natural resources also have a sacred and spiritual value to religious groups, indigenous peoples and those who value the cultural properties of the environment.

- Natural resources also have an intrinsic value – that is, the value a resource possesses as a result of being itself.

- This intrinsic value is often cited as a major reason for managing and conserving the environment, but it is not without its controversies.

Did you know?

Bake and shark is a famous dish from Trinidad and Tobago, consisting of fried shark meat in a fried dough 'sandwich'.

The ecosystem value of natural resources in the Caribbean

The valuation of ecosystems and the services they provide is the basis of the study of environmental economics. Because of the type and range of services that humans derive from nature, most of the values of these services are difficult to quantify, and they are often estimated. Examples of ecosystem services include many of the essentials of life: clean air and water; the gathering and production of food, fuel and raw materials from the land and sea; the regulation of climate; the attenuation of flooding and the filtration of pollutants, to name a few. Ecosystems, biodiversity and natural resources also give humans personal benefits from enjoying them that improve their health, happiness and well-being. Additionally, because the Caribbean relies heavily on tourism for its GDP, the aesthetic value of ecosystems, biodiversity and natural resources are fundamental, and perhaps priceless.

Caribbean states have to determine methods of valuing the region's ecosystem services, since this is crucial to understanding and recovering from events such as the 2005 floods in Guyana, the effects of Hurricanes Ivan and Emily in two consecutive years on Grenada, and the effect of the volcanic eruption of Soufrière Hills in Montserrat in 1995. Additionally, placing a 'price-tag' on natural resources may be a prudent way of fostering appreciation for ecosystems and the services they provide, and inspiring users to manage and conserve them sustainably. Guyana has valued its forest ecosystem as part of a bilateral agreement with the Republic of Norway, and in 2011 a second valuation was done of the reefs of the Caribbean and published in a report titled *Reefs at Risk Revisited*.

The role of natural resources in research and teaching

Research

Research on ecosystems and species is an important part of the management and conservation of ecosystems. This importance is largely reflected in the third objective of the 1992 Convention on Biodiversity: 'the fair and equitable sharing of benefits arising from genetic resources'. In 2010, the Conference of the Parties of the Convention adopted the Nagoya Protocol, whose objective is the fair and equitable sharing of benefits arising from the utilisation of genetic resources, thereby contributing to the conservation and sustainable use of biodiversity. The Protocol is meant to provide a transparent legal framework for the effective implementation of benefits arising out of the utilisation of genetic resources, and is exceedingly important for the Caribbean region given its high diversity of biodiversity and ecosystems, especially in the diverse coral reefs and rainforests.

Research on biodiversity and ecosystems is often done through bioprospecting – the discovery and commercialisation of new products based on biological resources. Bioprospecting often draws on indigenous knowledge about uses and characteristics of plants and animals, and has yielded new medicines such as quinine from the genus *Cinchona* as a result of oral traditions, and the discovery of new species – for example in 2012, two new species of orchid (*Tetramicra riparia* and *Encyclia navarroi*) were discovered in Cuba, and 24 new species of lizard were discovered by researchers from Penn State University in late 2012 in Anguilla, the British Virgin Islands, Jamaica and Dominica, among other islands. Bioprospecting has also given rise to biopiracy – the exploitative appropriation of traditional and indigenous forms of knowledge by commercial actors, as well as the search for previously unknown compounds in organisms that have never been used in traditional medicine, without the proper authorisation of a group of individuals or a state. While the Nagoya Protocol is meant to address these issues, it is not yet in force, and most Caribbean states do not have the legal framework to support the bioprospecting process.

As a result, indigenous knowledge of medicinal plants may be patented by companies, without recognising the fact that the knowledge is not new, or invented by the patenter, and depriving states and indigenous communities of the rights to commercial exploitation of technology that they had developed or practised. Often cited examples of biopiracy include the rosy periwinkle (*Catharanthus roseus*), the neem tree (*Azadirachta indica*), the enola bean, basmati rice and hoodia (*Hoodia* spp.). In 2000 the Mayan people of Chiapas, Mexico accused the International Cooperative Biodiversity Group of unethical forms of bioprospecting. This case is instructive to the Caribbean region, since it drew attention to the problems of distinguishing between bioprospecting and biopiracy, and to the difficulties of securing community participation and prior informed consent for bioprospectors.

Research is an important aspect of the education, awareness and advocacy process, and has been highlighted in the Tbilisi Declaration. Research may be conducted at the secondary level in the form of Internal Assessments, and is also a critical part of the work of tertiary institutions such as the University of the West Indies and the University of Guyana. Research by the Centre for Resource Management and Environmental Studies (CERMES) at the Cave Hill Campus of the University of the West Indies (UWI) in Barbados, the Bellairs Research Institute – a field-station for McGill University in Barbados, the International Centre for Biodiversity Research and Low Carbon Development of the University of Guyana and St George's University in Grenada are a few examples of research outfits which have contributed substantially to knowledge on marine biodiversity, fisheries, marine turtles, forests and terrestrial biodiversity and land-based sources of pollutants in the region.

Teaching

Teaching about natural resources, ecosystems, species and the environment is an important part of raising awareness of the value and threats of these entities. Programmes such as CAPE Environmental Science, and tertiary courses at the University of the West Indies, University of Guyana and University of Belize are examples of study programmes that focus on the environment.

Teaching may also be informal, in the form of education and awareness campaigns, radio and television programmes for example. Information can also be spread by social media, pamphlets and as part of plays and other cultural activities. Like research, teaching is an important aspect of the education, awareness and advocacy process, and has been highlighted in the Tbilisi Declaration (see 3.29).

Did you know?

Research can also be utilised to understand the habits and characteristics of particular species, especially those which are threatened, or about which little is known. Examples include the Nassau grouper (*Epinephelus striatus*) and the West Indian sea egg (*Tripneustes ventricosus*) – a species of sea urchin. In the wake of the invasion by the red lionfish (*Pterois volitans*) in Caribbean waters in recent times, there has been considerable research, as well as teaching, education and awareness-raising about the threat the species poses to the marine environment of the Caribbean region.

Links

See 3.29 for more on the Tbilisi Declaration.

Key points

■ The valuation of the region's biodiversity and ecosystem services is an essential aspect of appreciating the role and importance of these resources to the Caribbean people.

■ Research and teaching are important elements in recognising and valuing the region's biodiversity and ecosystems.

■ Research and teaching are also important elements in environmental education, advocacy and awareness.

On completion of this section, you should be able to:

■ identify political factors affecting government policies in the Caribbean region

■ explain how these political and economic factors affect natural resource use

■ outline environmental impact assessment as a natural resource management and conservation tool.

Factors affecting natural resource use

Political and economic factors are often key considerations which determine the way in which natural resources are utilised, and in what quantities. For countries in the Caribbean region where primary industries still dominate, there is often the impetus to exploit natural resources in order to try to make progress on the path to economic development.

Political factors

Political factors generally may shape natural resource use in terms of the policies, as well as the systems of regulation, adopted by the state. This includes policies which may offer incentives for the exploitation of natural resources and legislation.

Privatisation and private investment

When a government does not have the capital to invest in and sustain the exploitation of a resource, privatisation of an industry or private investment in resource exploitation can occur. This can lead to increased extraction of natural resources, since the private entity is principally concerned with profit. For this reason it needs to be adequately regulated by government to ensure compliance with environmental laws and policies. Examples include the granting of licences to private entities to exploit oil and natural gas (BPTT, Trinidad & Tobago), gold (Omai Gold Mines, Guyana) and bauxite (Alcoa Minerals, Jamaica).

Nationalisation

Nationalisation may refer to either direct state ownership and management of an enterprise, or the government acquiring a large controlling share of a corporation. An example of this type of acquisition is the bauxite industry in Guyana being nationalised in the 1970s.

Environmental and natural resources policies

In the Caribbean region natural resource policies include the exploitation of energy resources with emphasis on renewable bio- and agro-energy programmes, the management of marine resources and a framework for managing the Caribbean Sea (CARICOM, 2010). These policies can be aimed at previously unexploited resources such as geothermal, wave/tidal and ocean thermal energy conversion (OTEC). Other policies are aimed at existing natural resources such as oil/ natural gas and bauxite/ alumina, relevant to Trinidad & Tobago and Guyana. Policies which do not directly utilise natural resources, but rely on them indirectly, such as tourism and the offshore banking and finance industry is of relevance to the British Virgin Islands, Cayman Islands and St Kitts & Nevis.

Environmental impact assessments

An environmental impact assessment (EIA) is a formal process that the United Nations Environment Programme (UNEP) defines as 'an examination, analysis and assessment of planned activities with a view of ensuring environmentally sound and sustainable development' (UNEP, 1987). Principle 17 of the 1992 Rio Declaration on Environment and Development requires that the EIA as a national instrument 'be undertaken for proposed activities that are likely to have a significant adverse effect on the environment and are subject to a decision of a competent national authority'. EIAs began to be used in the 1960s as part of the decision making process, encompassing a technical evaluation of the positive or negative impacts that a proposed project may have on the environment.

The environmental impact assessment as it has evolved in the Caribbean encompasses several mechanisms, which further promote the ideals of sustainable development. Firstly, the nature of the process can promote the principle of public participation – in other words, the greater involvement of the public in decision making. This is because the EIA provides information, since it requires the collection of information about potential environmental impacts so that decision making is informed. Further, the process may yield information on alternatives that achieve the same developmental objectives, which may be less damaging to the environment. Information also provides the basis for deciding on any mitigation measures that may be appropriate. In this sense, the EIA is preventative, as it enables more environmentally sustainable decisions to be taken and therefore promote the principle of sustainable development.

Finally, the EIA process may also give rise to environmental justice, since citizens are afforded the opportunity to resort to courts or tribunals if they feel that their governments do not take sufficient consideration of the EIA process. This has been the case, to varying degrees of success in Belize, Trinidad and Tobago, the Bahamas, Jamaica and the Virgin Islands. There have been criticisms that there are inherent flaws in a largely procedural process, which mean that intended benefits of sustainable development may not be achieved. This is thought to be due to the lack of awareness, accountability and justice, and simply because there is not a culture of environmental advocacy in the Caribbean. However, this culture is changing, and the EIA will probably continue to feature prominently in the management and conservation of natural resources in the Caribbean region.

Key points

- Political factors include privatisation and nationalisation strategies, as well as environmental and natural resource policies.

3.13 Economic factors affecting natural resources use

Learning outcomes

On completion of this section, you should be able to:

- identify economic factors affecting government policies in the Caribbean region

- explain how these economic factors affect natural resource use.

 Links

See 3.28 for more about economic instruments.

Economic factors

Economic factors interact with the biophysical factors to shape natural resource use patterns. For example, resources which may be too remote, or in small concentrations, may not be economically viable given the market prices. Other factors include proximity to markets and buyers of resources.

Role of foreign investment

Similar issues apply to attracting foreign capital they do when there is any need for private investment and attention needs to be paid that this kind of investment does not lead to overuse of the resource or damage to the environment. Many states have begun to use economic instruments and mandatory restoration requirements a part of their economic and environmental policy. Examples of foreign investment in the use of natural resources include geothermal exploration in St Kitts & Nevis, ocean thermal energy conversion (OTEC) in the Bahamas.

Foreign investment needs to be adequately regulated by government to ensure compliance with environmental laws and policies, and to prevent environmental degradation

Export of natural resources as primary products

Caribbean states mainly export primary products, the prices of which are usually fixed on the world market, as a result, they often seek methods to increase or expand exploitation to make the industry profitable and to contribute substantially to GDP. Products such as bauxite, oil & gas and gold are exported in their primary form, however states such as Trinidad & Tobago, Jamaica and Barbados have manufacturing sectors, so export secondary products. Another secondary product most territories export is rum.

National debt

The need to service foreign debts can lead to overexploitation of natural resources. An example is Guyana, which at one time had such a high level of poverty and debt overhang that it qualified as a Heavily Indebted Poor Country (HIPC).

Sectoral activities

Tourism

Tourism is the main product of the Caribbean region with much direct investment going into the sector. The region places considerable emphasis on attracting investment to an industry that directly affects the coastal and marine ecosystems. Tourism is the mainstay of states such as Barbados. In the Bahamas, agencies establishing themselves in the tourism sector must pay a performance bond

Agriculture

Agriculture, in the form of sugar cane cultivation, was historically the main economic activity of the Caribbean region; however sugar cane cultivation has declined, and even ceased in some territories.

Many states in the OECS still plant crops such as bananas, and agricultural production still occurs in the larger territories such as Jamaica, Trinidad & Tobago and Guyana, and this has direct effects on the rivers, coastal and marine ecosystems.

Mining and manufacturing

Mining of various natural resources contributes significantly to the GDP of some Caribbean territories, but the activity affects the coastal and marine ecosystems thorough pollution and discharge of process chemicals. For this reason adequate regulation of activities needs to be in place. Examples mining include bauxite (Jamaica and Guyana), oil & natural gas (Trinidad & Tobago), limestone (Barbados), gravel (Dominica, Guyana) and gold (Guyana).

States such as Trinidad & Tobago, Jamaica and Barbados have manufacturing sectors, so export secondary products. This means in addition to the effects from the extraction of primary products, there are other effects associated with processing, such as pollution etc.. Examples manufacturing include cement by TCL, companies such as Grace (Jamaica), Pine Hill (Barbados) and Orchard (Trinidad & Tobago) which produce juices. Rum and spirits are produced in most Caribbean states and products derived from petroleum are manufactured in Trinidad & Tobago.

Forestry and fisheries

In addition to the effects from the harvesting of these primary products, there are other effects associated with processing. Forestry is present in states including Guyana, Suriname and Belize which export both primary and secondary products related to this industry.

The fisheries sector is fundamental to the majority of Caribbean states, but most of the catch is sold without processing to consumers, hotels or exporters. States including Guyana, Jamaica, Belize, Trinidad & Tobago process and export fish products.

Key points

- Several factors affect natural resource use in the Caribbean, both political and economic.

- Economic factors include national debt, the role of foreign investment and the export of natural resources as primary products.

- The environmental impact assessment (EIA) can be thought of as an important tool in adopting the tenets of sustainable development, and therefore including the concept in the management and conservation of natural resources.

- The EIA balances the social, economic and ecological aspects of a proposed project, and is usually regulated by legislation.

Did you know?

The EIA process was made a part of legislation in the United States in 1969 through the National Environmental Policy Act (NEPA), and since then it has evolved and is increasingly popular in many countries around the world. Within the Caribbean region, many states have legislation which expressly addresses the EIA – including Belize, St Kitts & Nevis, Trinidad & Tobago, Guyana and Jamaica. Barbados does not have specific legislation, but does require an EIA to be performed when planning permission is sought for developments on the coastal zone.

 Links

See 3.30–3.32 on environmental and conservation law.

Learning outcomes

On completion of this section, you should be able to:

- understand the concepts of species depletion and species extinction

- differentiate between species depletion and species extinction

- outline reasons for the occurrence of these events

- identify examples of these events in the Caribbean region.

Depletion, extinction and disruption

Three of the greatest threats to biodiversity are species depletion, species extinction and the disruption of habitats and ecosystems. The main causes for these effects vary from natural factors such as mass extinctions and co-extinction, to anthropogenic reasons such as overexploitation, urbanisation, pollution, climate change and the introduction of invasive species.

Species depletion and extinction

The species diversity and genetic diversity of the Caribbean region constitutes one of its most important assets. Unfortunately, there are many unsustainable practices which have led to large-scale species depletion and extinction in the region, including:

- The introduction of alien species – for example the white-tailed deer (*Odocoileus virginianus*), the small Asian mongoose (*Herpestes javanicus*) (which was first introduced in 1872 to control rodents and poisonous snakes on sugar cane plantations), and most recently the red lionfish (*Pterois volitans*) – has devastated native populations of reptiles and amphibians and marine life, and led to the extinction of dozens of species.

- Agriculture: the initial wave of forest clearing began in the early 1500s for sugar cane plantations, leading to widespread deforestation throughout the region. Sugar is still the Caribbean's most important crop. Agriculture remains a serious threat in parts of the region, with banana, cacao, coffee, and provision plantations threatening remaining large tracts of pristine forest, largely in the small island region (excluding Suriname, Guyana and Belize).

- Mining for bauxite, sand and gravel, as well as the production of charcoal from natural vegetation to meet energy needs, also pose threats to the native flora and fauna.

- Tourism has put pressure on natural ecosystems on some islands, particularly through the alteration of local landscapes with non-native vegetation, golf courses, roads, and tourist infrastructure and facilities. However, responsible tourism has emerged in some cases as a positive catalyst for conservation.

- Overharvesting and overexploitation of wildlife and fisheries for food, the wildlife trade, incidental loss due to habitat or ecosystem disruption or activities such as sport fishing.

- Pollution by land-based and marine sources originating from agriculture, industry, tourism and shipping. The main pollutant is sewage, but chemicals and hazardous wastes are also a significant threat.

Species depletion and extinction in the Caribbean region is exacerbated by endemism, since about one-quarter of the region's vascular flora is restricted to a single island – Cuba, and of the endemic genera in the region, about 120 are confined to single islands (e.g. the species of Amazon parrots in the Lesser Antilles) (Conservation International, 2012).

Did you know?

Vascular plants are plants that can transport water and food internally through special tissues in their roots, stems and leaves. The group of vascular plants includes most plants except for mosses, lichens, liverworts and algae.

∞ Links

See 3.3 for a definition of an endemic species.

Classification of species under the International Union for Conservation of Nature (IUCN) System

Species depletion will be reflected under the threatened categorisation, while extinction is considered the ultimate end of the spectrum. An important aspect of extinction at the present time is human attempts to preserve critically endangered species, which is reflected by the creation of the conservation status 'extinct in the wild' (EW). Species listed under this status by the IUCN are not known to have any living specimens in the wild, and are maintained only in zoos or other artificial environments. Some of these species are functionally extinct, as they are no longer part of their natural habitat and it is unlikely the species will ever be restored to the wild – for example Père David's Deer (*Elaphurus davidianus*), which has been extinct in the wild since 1865 (or possibly long before that) and the Pinta Island Tortoise (*Geochelone nigra abingdoni*), which had only one living individual, named Lonesome George, until his death in June 2012. The Pinta Island tortoise was believed to be entirely extinct in the mid-20th century, until Lonesome George was spotted on the Galapagos island of Pinta on 1 December 1971.

Another example is Spix's macaw (*Cyanopsitta spixii*), a species previously thought extinct. The IUCN regards it as critically endangered and possibly extinct in the wild. Its last known stronghold in the wild was in northeastern Bahia, Brazil and the last known wild bird was a male that vanished in 2000. The species is now maintained through a captive breeding programme at several conservation organisations under the aegis of the Brazilian government, is listed in Appendix I of CITES.

Comparing species depletion and species extinction

Species depletion will occur as a result of any of the factors outlined above, and if unchecked will lead to species extinction. Some examples of species depletion and species extinction are illustrated below (examples are illustrated in 3.4):

Species depletion is the disappearance of a species from part or all of its range. Species depletion will occur as a result of any of the factors outlined above, and if unchecked will lead to species extinction. A species becomes extinct when the last existing member dies. A species may become functionally extinct when only a handful of individuals survive, which cannot reproduce due to poor health, age or sparse distribution over a large area. Global extinctions are contrasted with local extinction, or extirpation, which occurs when a species (or other taxon) ceases to exist in the chosen geographic area of study, though it still exists elsewhere.

The extinction of one species' wild population can have knock-on effects, causing further extinction. These are also called `chains of extinction' and this is especially common with the extinction of a keystone species, and can lead to co-extinction. Co-extinction refers to the loss of a species due to the extinction of another. An example is the extinction of parasitic insects following the loss of their hosts. Co-extinction can also occur when a species loses its pollinator, or predators in a food chain lose their prey. Examples of co-extinction are Haast's Eagle (*Harpagornis moorei*) and its prey source the Moa (*Dinornis* spp.); this predator became extinct because its food source became extinct.

Did you know?

The dodo (*Raphus cucullatus*) of Mauritius, is an often cited example of modern extinction.

The passenger pigeon (*Ectopistes migratorius*) was hunted to extinction over the course of a few decades.

Stellar's sea-cow (*Hydrodamalis gigas*) was a large, herbivorous marine mammal, which was the largest member of the order Sirenia, which includes its closest living relative, the dugong (*Dugong dugon*), and the manatees (*Trichechus* spp.).

Key points

- Species depletion and extinction is one of the greatest threats to biodiversity in the Caribbean region.

- The main causes of these impacts include anthropogenic factors such as overexploitation, urbanisation, pollution, climate change and the introduction of invasive species, as well as natural factors.

- Species depletion and extinction in the Caribbean region is exacerbated by endemism, since many of the species in the region are confined to only one island.

Learning outcomes

On completion of this section, you should be able to:

- understand the concepts of habitat destruction and ecosystem destruction

- differentiate between habitat destruction and ecosystem destruction

- outline reasons for the occurrence of these events

- identify examples of these events in the Caribbean region.

Habitat disruption and destruction

Habitat disruption or degradation is interference or disturbance of a habitat by either natural or anthropogenic activities. Examples of natural activities include volcanism, fire, and climatic events such as hurricanes and tropical storms. However, at present, habitat disruption and destruction is the main anthropogenic cause of species extinctions. Examples of habitat degradation worldwide include:

- clearing habitats for agriculture (the principal cause of habitat disruption and destruction)
- harvesting natural resources for industry production
- urbanisation
- logging
- mining
- some fishing practices.

The degradation of a species' habitat may alter its ability to host a species to such an extent that the species is no longer able to survive, and will eventually become extinct. Degradation results from direct effects, such as pollution, or indirectly, by limiting a species' ability to compete effectively for diminished resources, or against new competitor species such as an invasive species. Global warming and climate change has also contributed to habitat degradation in the Caribbean – including the bleaching of coral reefs. Vital resources including water and food can also be limited during habitat degradation, leading to extinction.

Habitat destruction results from sustained and continued habitat degradation, and is a process of natural environmental change that may be caused by habitat fragmentation, which is the impact of discontinuities (fragmentation) on an organism's habitat, causing population fragmentation. Habitat fragmentation can be caused by geological processes that slowly alter the layout of the physical environment, or by human activity such as land conversion, which can alter the environment much faster and causes extinctions of many species. Geological processes, climate change or human activities such as the introduction of invasive species, ecosystem nutrient depletion and other human activities can also lead to habitat destruction. In this process, the organisms that previously inhabit the site are displaced or destroyed, reducing the species diversity of the area. Generally as a result of the process, the natural habitat is rendered functionally unable to support the species which it hosts. Habitats are destroyed mainly through the same mechanisms at play in habitat degradation.

Disruption of ecosystem processes

The Caribbean's ecosystems – forests, wetlands, mangroves, seagrass beds and coral reefs – have all been subject to disruptions. Much of the natural resources and ecosystems have been devastated on

⚭ Links

See Module 1 for ecosystem processes and 3.22 for tools of natural resource management.

some islands: for example, no more than 10 per cent of the original vegetation in the Caribbean islands remains in a pristine state. The significant environmental degradation began with the arrival of the first Europeans on Hispaniola in 1492, which triggered the initial wave of forest clearing in the early 1500s for sugar cane plantations. Sugar cane, which has led to widespread deforestation throughout the region, is also the source of other impacts of human settlers, most notably the introduction of alien species, which is the biggest threat to biodiversity in the region.

Forests continue to be destroyed for a variety of uses, such as logging, clearance for agriculture and fuelwood, but in many states consolidated efforts to protect them have been made by means of legislation (Jamaica, St Lucia), economic instruments (Guyana) and land management plans (Grenada). Wetlands, mangrove ecosystems and marine areas have also endured high levels of habitat destruction with marine coastal areas being highly modified by humans for urbanisation, to construct hotels, marinas, aquaculture facilities and in activities such as fishing and sand mining. Coral reefs and seagrass beds have also been destroyed or severely degraded by overfishing, pollution, and invasive species such as the red lionfish (*Pterois volitans*). The red lionfish are voracious feeders and have outcompeted and filled the niche of the overfished snapper and grouper, and as the fish become more abundant, they are becoming a threat to the fragile ecosystems they have invaded. Between outcompeting similar fish and having a varied diet, the lionfish is drastically changing and disrupting the food chains holding the marine ecosystems together. As these chains are disrupted, declining densities of other fish populations are found, as well as declines in the overall diversity of coral reef areas.

The disruption of the Caribbean's ecosystems leads to a reduction of ecosystem services, such as the attenuation of the force of waves by coral reefs, seagrass and mangrove ecosystems; the provision of food to species, as well as the availability of commercially fished species for humans. Processes such as the carbon, water and nitrogen cycles are disrupted, and the movement of energy through trophic levels will also be affected by shifts in the predator–prey dynamics.

Did you know?

The red lionfish (Figure 3.15.1) is a venomous coral reef fish native to coral reefs in the tropical waters of the South Pacific and Indian Oceans. Thought to have been introduced accidentally into the Caribbean region from an aquarium or similar facility, it has become a huge invasive problem in the Caribbean Sea and along the East Coast of the United States, along with a similar species, *Pterois miles* (the common lionfish or devil firefish).

Key points

- The disruption of habitats and ecosystems is one of the greatest threats to biodiversity in the Caribbean region.

- The main causes of these impacts include anthropogenic factors such as overexploitation, urbanisation, pollution, climate change and the introduction of invasive species.

- Habitat disruption may also lead to habitat fragmentation and eventually habitat destruction.

- This can set in train a sequence of events, including the disruption of ecosystem processes, species depletion and extinction.

Figure 3.15.1 *The red lionfish (Pterois volitans)*

Issues affecting water as a resource

Water is one of the most important natural resources, because it is required for all facets of basic human existence – direct consumption and domestic purposes, as well as for industry, agriculture, transportation, coolants and recreational purposes. However, threats such as the degradation, depletion and pollution of water threaten the availability of the resource for humans, and also expose them to health risks of waterborne diseases.

Pollution

Water pollution is the contamination of aquatic ecosystems: lakes, rivers, oceans, aquifers and groundwater by pollutants, which are discharged directly or indirectly into water bodies without adequate treatment to remove harmful compounds. Pollutants include any substance or energy introduced into the environment that has undesired effects, including heavy metals (e.g. mercury, lead, thallium and cadmium), pathogens, environmental pharmaceutical persistent pollutants (pharmaceuticals which linger and persist in the environment), persistent organic pollutants (POPs) – organic compounds that are resistant to environmental degradation through chemical, biological, and photolytic processes), and volatile organic compounds (organic chemicals that have a high vapour pressure at ordinary, room-temperature conditions).

Pathogens such as coliform bacteria are a commonly used bacterial indicator of water pollution: examples of organisms sometimes found in surface waters which have caused human health problems include *Giardia lamblia*, Salmonella, *E. coli* and parasitic worms.

Depletion and degradation of surface and groundwater resources

Water pollution affects all aquatic species and populations, as well as natural biological communities. For example, processes such as eutrophication affect the availability of oxygen for larger aquatic organisms such as fish, since they can be deprived of oxygen and die.

Eutrophication can lead to the stagnation of the waterway, and the waterway becoming hypoxic, while water contaminated by thermal pollution or sediments can affect the ability of aquatic plants to photosynthesise, and ultimately lead to a deterioration in the quality of water available for human consumption. This can lead to waterborne diseases such as schistosomiasis, typhoid, dysentery and even ear and skin infections.

Groundwater pollution occurs when a spill or ongoing releases of chemical or other contaminants seep into soil and into an aquifer or water located beneath the Earth's surface, creating a contaminant plume within an aquifer. Because groundwater is below the surface, its pollution is not as easily classified as surface water pollution, and often goes undetected until it emerges as a surface formation. Examples of incidences of groundwater contamination include the Love Canal incident in upstate New York and the Ganges Plain of northern India and Bangladesh, where severe contamination of groundwater by naturally occurring arsenic affects water wells in the shallower of two regional aquifers.

Seepage of pollutants into groundwater may be confined and therefore may not contaminate surface water or other features, but can contaminate the aquifer below. Movement of water and dispersion within the aquifer spreads the pollutant over a wider area, and also into surface water formations such as seeps and springs. This can make these water sources unsafe for humans and wildlife which use them.

Watershed destruction

The main sources of watershed destruction include:

- deforestation and destruction of natural vegetation for firewood extraction as well as arable and pastoral land; other major causes are wildfire and overexploitation
- overexploitation, which occurs when arable land is used beyond its fertility potential, which can lead to a variety of erosion features such as gully erosion, landslides and alternation of discharge
- overgrazing, which occurs when the number of livestock on an area of land is too large, leading to the destruction of natural vegetation as well as soil compaction and erosion: like overexploitation, overgrazing can result in various degradation features such as alternation of discharge, change of soil moisture and gully erosion
- unadjusted irrigation techniques: the implementation of certain irrigation techniques can lead to degradation primarily in the form of salinisation, for example, irrigated fields that are overwatered or badly drained can promote salinisation because these practices raise the water table and bring salts that have built up in the soil layers closer to the surface
- socioeconomic and political causes: population growth is a major cause of degradation, since it contributes to overgrazing, overexploitation and deforestation
- natural causes, which can be amplified through human action (such as pastoral and agricultural land use, for example): these include natural causes such as the nature of rainfall (amount, intensity, variability, distribution), soil (texture, structure, depth, moisture, infiltration rate). Topography also plays an important role in the scope and scale of degradation processes.

Key points

- Water is an essential natural resource for the Caribbean region.
- Water may be affected by many anthropogenic factors, including the physical conversion of land, pollution, discharge of process chemicals, sedimentation, siltation and oil spills.
- Human activities may also affect watersheds as well as surface and groundwater resources.

Did you know?

Watersheds are the areas that separate neighbouring drainage basins (catchments) or the areas of land where surface water from rain converges and joins another water body, such as a river, lake, reservoir, estuary, wetland, sea, or ocean.

 Links

See 3.5 for more on the importance of water in the Caribbean as well as the following section on water.

Learning outcomes

On completion of this section, you should be able to:

■ discuss the importance of water as a resource in the Caribbean region

■ identify the human health risks associated with improperly treated water.

Human health risks of improperly treated water

The human body contains from 55 to 78 per cent water, depending on body size, and requires between one and seven litres of water per day to avoid dehydration and function properly. Most of this is ingested through foods or beverages other than water. As a cardinal rule, humans require water with few impurities – known as potable water. Common impurities include metal salts and oxides, including copper, iron, calcium and lead and/or harmful bacteria. Potable water may be obtained by means of purification of rainwater, surface water or groundwater by filtration, distillation, or by a range of other methods. Some countries also utilise desalinisation to remove the salt and purify marine water.

The consumption of improperly treated water may result in waterborne diseases, which are typically caused by pathogenic microorganisms, and most commonly are transmitted in contaminated fresh water. High levels of pathogens may result from inadequately treated sewage discharges, caused by a sewage plant with a design that does not allow for secondary treatment of the sewage, as well as poorly managed livestock operations. Infection commonly results during bathing, washing, drinking, in the preparation of food, or the consumption of food thus infected.

Examples of waterborne diseases in the Caribbean:

Disease and transmission	Agent	Sources of agent in water supply	General symptoms
E. coli infection	Certain strains of *Escherichia coli* (commonly *E. coli*).	Water contaminated with the bacteria.	Mostly diarrhoea. Can cause death in immuno-compromised individuals, the very young and the elderly due to dehydration from prolonged illness.
Otitis externa (swimmer's ear)	Caused by a number of bacterial and fungal species.	Swimming in water contaminated by the responsible pathogens.	Ear canal swells, causing pain and tenderness to the touch.
Salmonellosis	Caused by many bacteria of genus *salmonella*.	Drinking water contaminated with the bacteria. More common in food.	Symptoms include diarrhoea, fever, vomiting and abdominal cramps.
Typhoid fever	*Salmonella typhi*.	Ingestion of water contaminated with faeces of an infected person.	Characterised by sustained fever up to 40 °C (104 °F), profuse sweating, diarrhoea may occur: symptoms progress to delirium and the spleen and liver enlarge if untreated (in this case it can last up to four weeks and cause death).
Dysentery	Caused by a number of species in the genera *Shigella* and *Salmonella* with the most common being *Shigella dysenteriae*.	Water contaminated with the bacterium.	Frequent passage of faeces with blood and/or mucus and in some cases vomiting of blood.

Schistosomiasis	Members of the genus *Schistosoma*.	Fresh water contaminated with certain types of snails that carry schistosomes.	Blood in urine (depending on the type of infection), rash or itchy skin, fever, chills, cough and muscle aches.
Hepatitis A	Hepatitis A virus (HAV).	Can manifest itself in water (and food).	Symptoms are only acute (i.e. there is no chronic stage to the virus) and include fatigue, fever, abdominal pain, nausea, diarrhoea, weight loss, itching, jaundice and depression.
Giardiasis	The protozoan *Giardia lamblia* is the most common intestinal parasite.	Untreated water, poor disinfection, pipe breaks, leaks, groundwater contamination, camping grounds where humans and wildlife use same source of water.	Diarrhoea, abdominal discomfort, bloating and flatulence.
Leptospirosis	Caused by bacterium of genus *Leptospira*.	Water contaminated by animal urine carrying the bacteria.	Begins with flu-like symptoms, with a second phase involving meningitis, liver damage (causes jaundice), and renal failure.
Cholera	Spread by the bacterium *Vibrio cholerae*.	Drinking water contaminated with the bacterium.	In severe forms it is known to be one of the most rapidly fatal illnesses known: symptoms include very watery diarrhoea, nausea, cramps, nosebleed, rapid pulse, vomiting, and hypovolemic shock (in severe cases), at which point death can occur in 12–18 hours.
Dengue, malaria and yellow fever	Mosquitos which transmit *Plasmodium* (in the case of malaria) and *Flaviviridae* (in the case of yellow fever).	Stagnant water.	Fever, headache, muscle and joint pains, and a characteristic skin rash that is similar to measles. Severe cases of malaria can progress to coma or death.

Key points

- Water has many domestic, industrial and recreational uses, and humans need to consume about one and seven litres of potable water per day to avoid dehydration and function properly.

- The populace of the Caribbean may be exposed to various health effects which arise from the consumption or the use of water for recreation.

∞ Links

See 3.5 for more on the importance of water in the Caribbean as well as the previous section on water.

On completion of this section, you should be able to:

- identify the major impacts of the extraction of minerals and hydrocarbons in the Caribbean region

- explain the reasons for these impacts.

⊖⊙ Links

See 3.17 for human health issues related to water resources in the Caribbean region.

The effect of exploiting minerals and hydrocarbons

Mining is the extraction of valuable minerals, hydrocarbons and other geological materials from the Earth. The extraction of these resources may take various forms – from excavating and stripping overburden in the case of bauxite, to the use of chemicals in the mining of gold, to the potential for the spillage and flow of harmful effluent in the case of hydrocarbon extraction. Environmental issues which arise from the process can include transformation of the landscape, erosion, loss of biodiversity, and contamination of soil, groundwater and surface water by chemicals from mining processes. In some cases, such as in Guyana, Suriname and Belize, areas of forests may be cleared in the vicinity of mines to increase accessibility to the resources. Some processes leave waste in the form of overburden (bauxite mining), slurry or tailing (gold mining), and techniques such as an illegal form of gold mining called missile dredging can damage the hydrology and dendrology of riverine ecosystems. Apart from environmental damage, contamination resulting from leakage of chemicals can also affect the health of the local population if not properly controlled. There have also been cases where mining activities or the transportation of hydrocarbons have resulted in fires which can last for long periods of time, causing massive environmental damage. Finally, safety during mining has long been a concern, and modern mining operations require occupational safety and health procedures in mines.

Generally the nature of mining processes creates a potential negative impact on the environment both during the mining operations and for years after the mine is closed. This impact has led to many states adopting regulations to moderate the negative effects of mining operations, including the posting of performance bonds, conducting environmental impact assessments and mandatory restoration at the end of the mining cycle.

Impact	Source	Effect
Physical conversion of vegetation and land	■ bauxite and gold mining. ■ mining gold and diamonds by land dredging.	■ overburden. ■ 'blue' lakes and sinkholes.
Transformation of landscape	■ bauxite and gold mining. ■ mining gold and diamonds by land dredging.	■ excavation of geographic features during open pit mining. ■ overburden. ■ 'blue' lakes and sinkholes.
Dust and noise pollution	■ sand, gravel, bauxite, mineral and hydrocarbon extraction, limestone.	■ dust from extraction of overburden in bauxite mining. ■ noise from machinery. ■ dust from processing plants: smelting to alumina and aluminium. ■ dust from the crushing of gravel and limestone. ■ dust pollution if vehicles transporting sand are not properly covered during transport.

Pollution from the discharge of process chemicals	■ bauxite and gold mining.	■ mercury and cyanide from small/ medium and large-scale gold mining respectively. ■ ferruginous residue (red mud) from bauxite processing. ■ 'tailings' (waste) and slurry from gold.
Sedimentation and siltation	■ sand, gravel, bauxite, gold, mineral and, limestone.	■ tailings ponds (impoundments, dams or embankment dams) in gold mining. ■ 'blue' lakes and sinkholes in bauxite mining.
Beach loss and the change in river course	■ sand mining and gold mining. ■ mining gold and diamonds by land dredging.	■ removal of sand from beaches (now illegal in most Caribbean states). ■ missile dredging – a diver-less hose is used to suck up unconsolidated gold-bearing gravel from the river bed, or blast the river banks for gold-bearing veins (illegal in most states, but practised because of the efficiency with which the ore may be collected).
Oil spills	■ hydrocarbon extraction and transportation.	■ fires and spills during extraction and transportation. ■ 'flaring' off the gas (though this is not as widely practised as in the past, since natural gas is now utilised).
Human health risks	■ sand, gravel, bauxite, gold, mineral and hydrocarbon extraction, limestone.	■ safety from mine collapses, fire and accidents related to machinery. ■ waterborne and other vector borne diseases such as malaria, typhoid and dengue fever. ■ sexually transmitted diseases. ■ heavy metal poisoning from mercury in gold mining.
Social dynamics (displacement of communities and the introduction of new settlements)	■ bauxite, gold, mineral and hydrocarbon extraction.	■ movement of populations to areas where there is a mining rush. ■ new towns and settlements being established. ■ 'absent male' syndrome. ■ prostitution, gambling, etc., in mining areas.

Figure 3.18.1 Open pit mining

Key points

- The exploitation of minerals and hydrocarbons has had significant effects on the ecosystems and biodiversity in the Caribbean region.

- These effects include the physical conversion of land, transformation of the landscape, pollution, discharge of process chemicals, sedimentation, siltation, oil spills, beach loss and the change in river courses.

- These industries can also have impacts on the health of the populace, as well as altering their social dynamics.

Soil, landscape and seascape transformation

Soil, landscape and seascapes are important aspects of the Caribbean identity, with many of the coastal ecosystems such as mangroves, seagrasses and coral reefs providing essential services and livelihood to the region's populace. However, these features are impacted by a variety of sources, which can threaten their viability and existence.

Soil, landscape and beaches

The natural landscape is transformed to the built environment through urbanisation and the construction of hotels, marinas and aquaculture facilities. This can cause erosion of coastlines, interfere with depositional processes, pollution of the coastal zone and loss of beach and coastline. Beach erosion can be caused by sand mining and removal of coastal vegetation, which again causes interference to depositional processes and loss of coastline. Activities such as banana cultivation and clearing slopes for agricultural purposes can lead to soil degradation such as erosion, landslides, desertification and also cause siltation of surface water features such as rivers lakes and streams.

Natural resource use also impacts the seascape in various ways, and examples of the effects are outlined in the following boxes about coral reefs, seagrass beds and mangroves.

Coral reefs

Coral reefs, which form some of the most diverse ecosystems on Earth, are underwater structures made from calcium carbonate secreted by corals. They are found in warm marine waters that contain few nutrients and permit the penetration of light. They provide ecosystem services and have a positive effect on tourism, fisheries and offer shoreline protection. They are fragile ecosystems, partly because they are very sensitive to water temperature and sedimentation. Activities which more obviously and directly threaten coral reefs are coral mining and the digging of marinas and access into islands and bays. Their fragile nature also makes them vulnerable to many things, including agricultural and urban runoff, pollution (organic and inorganic), disease, sea level and temperature rises pH changes from ocean acidification (all associated with greenhouse gas emissions). Certain fishing methods such as blast fishing and the use of cyanide are a threat as is fishing for aquariums and simple overfishing. Death and loss of coral reefs and their functions, habitat and species can result from these activities.

Seagrass beds

These consist of flowering plants from one of four plant families which grow in marine, fully saline environments; they are found in warm marine waters that contain few nutrients and permit the penetration of light. Sometimes labelled `ecosystem engineers', as they partly create their own habitat. The leaves of seagrasses slow down water currents, thus increasing sedimentation, and the roots and rhizomes stabilise the seabed. They provide important habitats for associated species mainly due to provision of shelter (through their three dimensional structure in the water column), and for their extraordinarily high rate of primary production. Seagrasses provide coastal zones with a number of ecosystem goods and ecosystem services, for instance fishing grounds, wave protection, oxygen production and protection against coastal erosion. Seagrass meadows account for 15% of the ocean's total carbon storage, and are therefore an important link in the carbon cycle (Module 1): per hectare, seagrass meadows hold twice as much carbon dioxide as rainforests (UNEP, 2009). Yearly, seagrasses sequester about 27.4 million tons of CO_2, but due to global warming, some seagrasses will go extinct (e.g. *Posidonia oceanica*, which is anticipated to go extinct, or nearly so, by 2050). These are fragile ecosystems, partly because they are very sensitive to water temperature and sedimentation. Eutrophication (excessive input of nutrients; nitrogen and phosphorus) is directly toxic to seagrasses, but most importantly, it stimulates the growth of algae, which reduces penetration of sunlight and the ability of the grasses to photosynthesise. Other threats include mechanical destruction for construction of marinas, aquaculture facilities. Loss of the functions of the seagrass meadows and their habitat and associated species has a negative impact on the environment.

Mangroves

Mangroves are various species of trees which grow in saline coastal sediment habitats in the tropics. Mangroves may be categorised as a wetland species and an ecotone, and the ecosystem is a distinct saline woodland habitat. The habitat is characterised by depositional coastal environments, where fine sediments (often with high organic content) collect in areas protected from highenergy wave action. The saline conditions tolerated by various mangrove species range from brackish water, through pure seawater (i.e. 30 to 40 ppt) to water concentrated by evaporation to over twice the salinity of ocean seawater (up to 90 ppt). The main species common to the Caribbean region are red mangrove (*Rhizophora mangle*), black mangrove (*Avicenna germinans*) and white mangrove (*Laguncularia racemosa*), with the mangrove associate buttonwood (*Conocarpus erectus*), often associated with the ecosystem.

Mangrove swamps protect coastal areas from erosion and storm surge, especially during hurricanes and tsunamis. The massive root systems of mangroves are efficient at dissipating wave energy, and slowing down tidal water enough so its sediment is deposited as the tide comes in, leaving all except fine particles when the tide ebbs. The unique ecosystem found in the intricate mesh of mangrove roots offers a quiet marine region for young organisms. Mangroves also host several commercially important species of fishes and crustaceans. In areas where roots are permanently submerged, the organisms they host include a wide variety of marine and freshwater organisms, which all require a hard surface for anchoring while they filter feed. Mangroves are often the object of conservation programmes, including national biodiversity action plans due to the specialityof mangrove ecosystems and the protection against erosion they provide.

Mangroves dominate many tropical coastlines, but their numbers have been declining for a variety of reasons, including urbanisation and other construction. This specific ecotone is also subject to many of the same threats as seagrass beds and coral reefs.

 Links

See 3.7 for human health issues related to water resources in the Caribbean region.

Key points

- Soil, landscape and seascapes are important aspects of the Caribbean identity.

- Mangroves, seagrasses and coral reefs provide essential services and livelihood to the region's populace.

- These features are impacted by a variety of sources, which can threaten their viability and existence.

3.20 The conservation and management of natural resources

Learning outcomes

On completion of this section, you should be able to:

- identify the measures used for the management and conservation of natural resources in the Caribbean region

- understand the fundamentals of the strategies for the management and conservation of natural resources in the Caribbean region.

Did you know?

The term 'conservation' came into use in the late 19th century, and refers to the management of valuable natural resources such as timber, fish, water and minerals. As the concept of sustainable development evolved, conservation expanded to encompass the preservation of forests and other ecosystems, cultural resources and wildlife. Conservation of natural resources embraces the broader conception of conserving the Earth itself by protecting its capacity to regenerate and recover. Particularly complex are the problems of non-renewable resources such as oil and other minerals in great demand.

Management

The principal objective of natural resource management is directing the way in which people and natural landscapes interact. It brings together facets such as land use planning, water management, biodiversity conservation, and the future sustainability of industries like agriculture, mining, tourism, fisheries and forestry. People and their livelihoods rely on the health and productivity of natural resources and ecosystems, and their patterns of use and consumerism play a critical role in maintaining the health and productivity of the environment.

Rehabilitation and restoration

The term 'restoration' in its most formal sense means returning an ecosystem to its original pre-disturbance state. The reasons for restoration include preventing habitat loss and fragmentation especially for endangered species. It can also be approached from an ethical and spiritual perspective human considering that beings have degraded, and sometimes destroyed ecosystems, so it is their responsibility to 'fix' it. Restorations can be restorations expensive and therefore not economically feasible and also, sadly, they don't always work. It could also be considered that by means of secondary succession, ecosystems may recover by themselves. Examples of restoration steps are removal of obstructions such as hydropower dams in rivers and removal of feeder roads, trails and other points of access in an ecosystem.

Rehabilitation refers to all types of habitat manipulations including enhancement, improvement, mitigation, habitat creation, and other situations, which do not truly restore a system, but restore specific aspects of the system. It shares similar purposes to restoration in terms of habitats, promoting secondary succession and halting disturbances, but has a primary aim of restoring productivity. Examples include mine rehabilitation, constructed wetlands to regenerate a farmed or forested area and reforestation. Restoration is more complex and in many areas where the land use is predominantly agricultural, residential, urban or industrial, true restoration is not feasible (Stanford *et al.*, 1996).

Preservation

Ecosystem preservation is perhaps the strictest form of natural resource conservation. Preservation is usually achieved by using legal means to ensure that the ecosystem is unavailable to development or exploitation by humans. An example is the designation of a protected area under Category1a of the IUCN scheme. The main challenge with this approach is that many of the natural resources and ecosystems that are threatened are also required by humans for their everyday needs. In many Caribbean states, where there is limited space and resources, this is often an impractical option. This measure should therefore be reserved for ecosystems that are in dire and immediate need of protection, as opposed to ecosystems which may be managed by other means.

In-situ and ex-situ conservation

Protecting an endangered plant or animal species in its natural habitat (in-situ), is achieved by protecting or cleaning up the habitat itself, or by defending the species from predators. This maintains recovering populations in the surroundings where they have developed their distinctive properties. The positive effect of this is it allows the ongoing processes of evolution and adaptation within their environments. In-situ conservation need to be of sufficient size to enable and maintain the target species to a desired number , but conflicting land use priorities be mitigate by zoning or the use of multi-use protected areas. Examples of in-situ approaches are wildlife corridors and agroecosystems. These may be used in tandem with ex-situ conservation, e.g. captive breeding programmes.

Ex-situ conservation consists of moving a the threatened species, or part of the population, from a fragmented or damaged habitat to a regulated environment. There can be scientific and other reasons for this too such as preserving genes or species for wider conservation purposes. Ex-situ is less preferable to in-situ and should be used as a last resort, or as a supplement to in-situ conservation because it cannot recreate the natural habitat fully. Natural evolution and adaptation processes are either temporarily halted or altered. Other methods include captive breeding programmes, seed and gene banks and cryogenic storage methods. However certain plant genera with recalcitrant seeds do not remain fertile for long periods of time, and a species grown in an artificial environment may not develop a resistance to diseases and pests foreign to the species, to which it has no natural defence.

Any approach needs to balance the ecological, social and also economic needs of natural resources. Balancing of the competing uses may prove challenging with competing land uses, especially on the coastal zone.

Key points

- Natural resource conservation involves the management of valuable natural resources, both renewable and non-renewable.

- Where the natural resource – especially a renewable resource – is in a dire state, the most prohibitive form of management, preservation, is used.

- Rehabilitation and restoration are usually concerned with returning the environment to a functional or original state, after the extraction and use of resources such as timber, gold or other minerals.

Did you know?

In-situ conservation or 'on-site' conservation is the promotion of conservation measures for species, ecosystems and their services as they occur naturally, while ex-situ or 'off-site conservation' is the process of protecting an endangered species of plant or animal outside its natural habitat. Collectively, these measures are considered a central part of the protection of biodiversity under the 1992 Convention on Biological Diversity.

Links

See 3.25 for a further discussion of protected areas as in-situ conservation in the Caribbean region, and 3.30 for the role of the 1992 Convention on Biological Diversity in in-situ conservation.

Reasons for the conservation and management of natural resources

On completion of this section, you should be able to:

- identify the reasons for the management and conservation of natural resources in the Caribbean region
- understand the fundamental theories and principles underpinning the management and conservation of natural resources in the Caribbean region.

Why conserve natural resources?

People conserve natural resources for a number of reasons. Broadly, these reasons may be categorised into ecological, economic and ethical reasons.

Ecological reasons

The pressures of population, technology and various human activities have had numerous effects on the functioning of ecosystems and species. These include habitat destruction and fragmentation, loss of biodiversity resulting from the introduction of invasive species and species extinction, and pollution – to name just a few. A turning point in people's attitude is recognising that the effects they have on ecosystems in turn affect their own lives, health and livelihood. The threats of habitat loss, species depletion and species extinction, along with the services provided by the Caribbean's ecosystems, are the key ecological reasons for the conservation of natural resources. Examples include:

- loss of ecosystems such as coral reefs, seagrass beds and mangroves, which affect ecosystem productivity, fisheries, tourism and coastal stability
- the need to maintain ecosystem functions such as the water, carbon, nitrogen and phosphorus cycles
- global warming and climate change, which is directly linked to the carbon cycle
- ensuring viable supplies of fresh water, which is linked to both the water and carbon cycles, as well as to pollution
- prevention of the loss of wild and commercial species (e.g. commercially valuable fish)
- maintaining genetic diversity by retaining wild populations.

Economic reasons

Economic impact studies document the many and substantial economic benefits generated by biodiversity. These include:

- provision of wild harvested food products such as fish, large and small animals, and non-timber forest products (NTFPs)
- optimal functioning of the agricultural and forest industry through processes such as pollination, pest control, nutrient provision, genetic diversity, and disease prevention and control
- provision of medicinal plants and raw materials for pharmaceuticals
- enabling nature-based tourism and the hunting and fishing industry
- provision of ecosystem services such as shoreline stabilisation and carbon sequestration.

Ethical, sacred and spiritual reasons

Ethical reasons for the conservation of the environment centre on the responsibilities that human have towards wild species and ecosystems — and to present and future generations of humans dependent on critical ecological services. Environmental ethics is a branch of applied philosophy that studies the conceptual foundations of environmental values as well as more concrete issues surrounding societal attitudes, actions, and policies to protect and sustain biodiversity and ecological systems. There are differing views on this subject, held by anthropocentrists and non-anthropocentrists respectively. Anthropocentrists believe that the environment's usefulness to humans is the most important consideration when it comes to conservation, while the non-anthropocentrists respect species and ecosystems for their own sake and not because of their utility to mankind.

Since the 1960s there has been growing interest in 'spiritual ecology', and this interest has accelerated markedly since the 1990s with an increasing number of conferences and publications. Spiritual ecology focuses on the relationships between religions and environments from the local to the global levels to address environmental crises, problems and issues. One component of spiritual ecology is sacred places – places that are special because of some extraordinary attributes that stimulate feelings of power, mystery, awe, transcendence, peace and healing. These include areas used for religious purposes, such as the Caroni River in Trinidad, as well as sites that may have some cultural and historical significance, e.g. Sauteurs in Grenada.

Aesthetic value

For the Caribbean region, the image of sun, sand and sea is the basis of the ecotourism industry, which is one of the major GDP earners in the region. Tourism may be nature-based as in the case of Dominica, Suriname and Guyana, or traditional as the case of most of the Caribbean, but the fundamental requirement is the ecosystems, species and natural resources of the region. Tourists come to zipline through forests, lie on the beaches and snorkel with turtles, so if these features are no longer there, tourists will have no incentive to visit.

Key points

- Mankind is heavily dependent on natural resources for a wide range of reasons – including ecological, economic, aesthetic and ethical, sacred and spiritual reasons.

- Ecological reasons are associated with ecosystem services such as the hydrological and carbon cycles, the management of habitat loss and species depletion/extinction.

- Economic reasons stem from the contribution of natural resources to the GDP and the livelihoods of the populace of the region.

- Ethical, sacred and spiritual reasons concern the proposition that mankind has a responsibility to wildlife and ecosystems, and that some members of the population – e.g. indigenous people, or a certain religious group – may hold the environment sacred.

- Aesthetic reasons are important both for the use of the resources by the population and for the tourism industry, which is one of the main contributors to GDP in the Caribbean.

 Links

See 3.33 for more information about the historic importance of Sauteurs.

Links

See 3.25 for a further discussion of protected areas as in-situ conservation in the Caribbean region, and 3.30 for the role of the 1992 Convention on Biological Diversity in in-situ conservation.

3.22 Tools for the management of renewable and non-renewable natural resources

Learning outcomes

On completion of this section, you should be able to:

- identify measures for the management of renewable and non-renewable resources in the Caribbean region

- distinguish between these measures for the management of renewable and non-renewable resources, based on their characteristics.

Tools and measures for natural resource conservation

Natural resources are desired for human use and consumption. In order to ensure their sustainable utilisation, many tools and measures are used in natural resource management and conservation. Because of the nature of renewable as opposed to non-renewable resources, tools often differ in their method of implementation, while some measures are aimed at influencing the human attitude to the use of the resources.

Tools and techniques for renewable resources

Sustainable yield management

Renewable resources are generally able to regenerate themselves in short cycle time, and hence are available for use by mankind. However, with the increase in consumerism, renewable resources are being consumed at rates which far exceed their ability to regenerate themselves. The concept of the maximum sustainable yield (MSY) refers to the largest yield (or catch) that can be taken from a species' stock over an indefinite period of time without the stock becoming unviable.

The concept of MSY is used extensively in fisheries, where it reflects the largest quota which may be harvested from a fishery stock over a period. Generally, the MSY will be exactly half the carrying capacity of a species, because at this yield level subsequent population growth will be highest. The concept may also be applied to the harvesting and use of forest resources, which are another renewable resource which is in high demand by humans.

The approach is not without its challenges, because most models are concerned with rates of extraction in relation to the rate of reproduction of a specific species, while ignoring other factors that may affect the resource. Such factors include the damage to the ecosystem caused by pollution, urbanisation and other coastal impacts, and the issue of fish caught unintentionally while intending to catch other fish (which may be undersized individuals or juveniles of the target or non-target species).

Tools and techniques for non-renewable resources

Extraction of natural resources

The conservation of non-renewable resources faces additional challenges because they are present in fixed quantities that diminish over time as they are exploited. As a result, extraction and use of non-renewable

Did you know?

Maximum sustainable yield may also be used to manage forestry resources.

resources often presents a challenge to the paradigm of sustainable development, since, once extracted, these resources cannot be used by future generations. The focus of sustainable management of these resources therefore has to be on finding alternative approaches to the consumption of the resources, based on a just distribution of benefits between today's consumers and those of future generations.

Use of substitutes and appropriate technology

Substitute fossil fuels include alternative sources of energy such as solar energy, wind energy, tidal energy, energy from biomass (biogas), etc. These sources of energy have not until now been used on a large scale, but their use is increasing. In some Amerindian villages in Guyana, solar cookers and biogas are now used for cooking and new materials are being developed to substitute for non-renewable resources. For example, plastics are now used to make products that once could be made only out of steel.

The movement toward the use of substitutes and appropriate technology arose out of a combination of economic and ecological considerations. The appropriate technology movement was precipitated by the energy crisis of the 1970s; it focuses on technological solutions to the energy crisis, and especially on ways of combating poverty by using cheap appropriate technology in the developing world. Appropriate technology depends on the characteristics and needs of the place where it will be applied, partly because by definition it involves the idea that a technology 'appropriate' to one set of circumstances is not necessarily appropriate for another. Nevertheless, some consistent features include technology that is simple to apply, not capital- or energy-intensive – i.e. requires little non-renewable energy to build, run or maintain, uses local resources and labour; and nurtures the environment and human health. Examples of appropriate technology in the Caribbean include photovoltaic cells and biofuels.

 Links

See 3.2 for the major categories, location and distribution of the major kinds of natural resources in the Caribbean.

Key points

- The maximum sustainable yield (MSY) is a tool used for the management of renewable resources.

- Its aim is to maintain a fisheries at a sustainable level, by reflecting the largest quota which may be harvested from a fishery stock over a period, without depleting the resource.

- It is also applicable to managing forestry resources.

- The use of appropriate technology involves the idea of utilising alternative technology that is particularly suited to the needs of a geographic location.

On completion of this section, you should be able to:

■ identify ways of reducing and minimising waste

■ identify ways of recycling, repairing and re-using materials.

Reduction and minimisation of waste

The waste hierarchy is a classification of waste management options in order of their environmental impact, such as reduction, re-use, recycling and recovery. In Europe the waste hierarchy has five steps:

1 prevention;

2 preparing for re-use;

3 recycling;

4 other recovery, e.g. energy recovery; and

5 disposal.

In the waste hierarchy, the most effective approaches to managing waste are at the top. In contrast to waste minimisation, waste management focuses on processing waste after it is created, concentrating on re-use, recycling, and waste-to-energy conversion. Waste minimisation refers to the reduction of the production of waste at society and individual level. Waste minimisation usually requires knowledge of the production process, cradle-to-grave analysis – the tracking of materials from their extraction to their return to Earth – and detailed knowledge of the composition of the waste.

Waste minimisation may have high initial capital costs, and many governments have introduced incentives for waste reduction. Examples of waste minimisation include:

■ re-using scrap material: this is the process by which individuals and industry re-use the waste materials they produce

■ exchanging waste: this is the technique in which the waste product of one process becomes a raw material for another process

■ minimising or eliminating product packaging

■ durability: improving product durability is another way to reduce waste as people do not have to replace them so often

■ changes in the style of living: waste is produced when new products are bought even though you have already useable but older products.

■ home composting using garden and kitchen waste

■ buying products with longer lives, and mending and repairing broken or damaged products.

Recycling, repair and re-use

All types of metal wastes, glass and paper and plastic may be recycled and used again. This includes paper and metal articles, as well as items made from plastic, which is largely derived from petroleum. Recycling plastic helps to conserve fossil fuel, while recycling paper helps to conserve forests. Another metal which is frequently recycled is aluminium, which involves re-melting the metal from scrap or waste materials. Recycling of aluminium is particularly advantageous, since the process is far less expensive and energy-intensive than creating new aluminium through the electrolysis of aluminium oxide, which must first be mined from bauxite ore and then refined. Additionally, bauxite is a non-renewable resource, which mean that there are finite known reserves of the mineral at present. Used beverage containers are the largest component of processed aluminium scrap.

Did you know?

The recycling of aluminium generally produces significant cost savings over the production of new aluminium, even when the cost of collection, separation and recycling are taken into account. Over the long term, even larger national savings are made through the reduction in the capital costs associated with landfills. In addition, the emissions from smelting plants using fossil fuels may be reduced from the recycling process.

Figure 3.23.1 *Aluminium cans*

Another conservation strategy is to repair and re-use materials – for example domestic and office appliances – instead of consuming new resources. This helps to conserve resources as it discourages production and wastage. However, this has to be balanced against side-effects – for example the potential to produce pollution by running out-of-date equipment that may use excessive amounts of fuel compared to the new equivalent.

Did you know?

Using public transport instead of individual vehicles reduces the use of fossil fuels. The 'car-pool' system, where several people with a common destination travel in one vehicle, also saves fuel. Additionally, switching off fans, lights and coolers when not in use, and using fluorescent rather than traditional electric bulbs, are some ways of conserving non-renewable resources.

 Links

See the previous section on tools for the management of renewable and non-renewable natural resources.

Key points

■ Waste minimisation refers to the reduction of the production of waste at society and individual level, and usually requires knowledge of the production process, cradle-to-grave analysis and detailed knowledge of the composition of the waste.

■ Recycling involves the processing of waste materials into new, usable products to prevent waste of potentially useful materials.

■ Recycling also has advantages such as the reduction in the consumption of fresh raw materials, the reduction of energy usage, reduction of pollution (air and water) and lower greenhouse gas emissions as compared to production processes for new products.

On completion of this section, you should be able to:

- explain the use of coastal and land use planning and zoning for natural resource management and conservation in the Caribbean region.

The majority of the population in the Caribbean live in the coastal zone. This creates land use and resource issues, many of which are related to urbanisation, tourism development (hotels, marinas, etc.) and aquaculture.

Land use planning and zoning

One of the oldest mechanisms used in the Caribbean to address the management and use of natural resources and the environment is land use planning and zoning. Most of the planning regimes were transplanted into Caribbean states after the Second World War from the colonial power, where they had evolved mainly in response to the expansion and ill-effects on the health of the populace after the Industrial Revolution.

Land use planning encompasses those measures adopted by a state to regulate land use in a balanced and effective way so as to prevent conflicts in land use. Accordingly, planning entails providing for the needs of the community while safeguarding natural resources. The principal tool used in planning is to first assess the characteristics of land, and planners often seek to combine and regulate compatible uses by the practice of zoning. As a tool for implementing land use plans, zoning regulates the types of activities that can be accommodated on a given piece of land, the amount of space devoted to those activities, and the ways that buildings may be placed and shaped. This is historically achieved through land use regulations.

This is especially crucial in the majority of Caribbean states, which have seen increased levels of urbanisation, tourism developments such as hotels and marinas and the use of coastal areas for agriculture and aquaculture. Planners and citizens often take on an advocacy role during the planning process in an attempt to influence public policy, but often, due to political and economic factors, governments are slow to adopt land use policies that are congruent with scientific data supporting more environmentally sensitive regulations.

Land use planning and zoning has evolved from its initial objectives in the colonial era to encompass a wider gamut of contemporary problems. This has included its expansion into environmental planning, which takes in many of the ecological and social implications of development actions.

Coastal zone planning

The increasing importance of the coastal region to the Caribbean hinges on the fact that approximately three quarters of the population of the region live within 200 km of the coast (Hinrichsen, 1998). A general cross section of the coastal zone takes in terrestrial ecosystems, beaches, wetlands, mangroves, seagrass beds and coral reefs. Definitions of coastal zones therefore tend to reflect the emphasis placed by different user groups on the various resources, including space, contained within coastal areas.

The key environmental problems facing the coastal and marine areas of the Caribbean are related to habitat conversion and destruction, pollution produced by human activities, and overexploitation of coastal and offshore fisheries resources. The underlying causes of these problems are linked to the ongoing development of coastal areas for tourism, infrastructure and urbanisation, and to the conversion of coastal habitats for uses such as agriculture and the expansion of aquaculture. In addition to diminished natural productivity of coastal areas, there has been a general lack of effective coastal regulations and enforcement (Caribbean Environmental Outlook, 2005). These factors threaten to undermine the delivery of critical ecosystem services, including the provision of fisheries, places for recreation and tourism, and control of pests and pathogens, and are expected to be significantly exacerbated by climate change and sea-level rise.

Integrated coastal management (ICM) can be thought of as the continuous and dynamic process by which decisions are made for sustainable use, development and protection of coastal marine areas and resources. Efforts in this regard were thought to be catalysed by, and grew exponentially as a result of, the 1992 Rio Conference, and in particular Chapter 17 of Agenda 21. One of the major catalysts behind ICM is to overcome the fragmentation inherent in both the sectoral and management approach and the splits in jurisdiction among levels of government at the land–water interface. ICM now encompasses a comprehensive framework which does not solely consider the land–water interface, but also the processes and factors that have a cause–effect relationship on the coastal area.

The transitional nature of coastal features which lie between the ocean and land environments produces changes in the physical environment across the coastal zone. These include a transition of ecosystems, which in turn result in successive ecotones, often associated with of different species occurring in each ecotone.

 Links

See 3.29 on environmental awareness and advocacy.

Key points

- Land use planning is among the oldest methods for the management and use of natural resources and the environment.

- Its main purpose is to regulate the balanced mix uses of any sector of land with respect to its capacity for supporting human development and use, animal, and plant life in harmony, since upsetting this balance has dire consequences on the environment.

- With the increased levels of urbanisation, tourism developments such as hotels and marinas, and the use of coastal areas for agriculture and aquaculture, the Caribbean region has incorporated integrated coastal management (ICM) as a tool for natural resource management.

- ICM encompasses a comprehensive framework which does not solely consider the land–water interface, but also the processes and factors that have a cause–effect relationship on the coastal area.

3.25 Protected area systems for natural resource management

Learning outcomes

On completion of this section, you should be able to:

- identify the features of protected area management in natural resource management and conservation in the Caribbean region

- explain the role of protected area systems as a tool for natural resource management and conservation in the Caribbean region.

Did you know?

The International Union for Conservation of Nature (IUCN) is currently the most prominent international organisation working on the subject of protected areas and marine protected areas (MPAs). The methodology utilised by the IUCN has become the international blueprint for designing a protected area management regime, and is the principal framework used in protected area management in the Caribbean region.

Protected areas

The definition of 'protected areas' that is most widely accepted is that given by the IUCN in its categorisation guidelines for protected areas. The definition is as follows:

> A clearly defined geographical space, recognised, dedicated and managed, through legal or other effective means, to achieve the long-term conservation of nature with associated ecosystem services and cultural values.

This definition applies to any protected area – whether terrestrial, marine or transboundary, and regardless of the management objectives. The IUCN has developed the protected area management categories system to define, record and classify the wide variety of specific aims and concerns that arise when categorising protected areas and their objectives:

Category		Description
1	Category Ia – Strict Nature Reserve	Area which is protected from all but light human use in order to preserve the geological and geomorphical features of the region and its biodiversity. If perpetual intervention is required to maintain these strict guidelines, the area will often fall into category IV or V.
	Category Ib – Wilderness Area	Areas generally larger and protected in a slightly less stringent manner than in the case of Strict Nature Reserves. These areas are protected domain in which biodiversity and ecosystem processes are allowed to flourish or experience restoration if previously disturbed by human activity. These are areas which may buffer against the effects of climate change and protect threatened species and ecological communities.
Category II – National Park		This bears similar characteristics to Wilderness Areas with regard to size and the main objective of protecting functioning ecosystems. National Parks tend to be more lenient in terms of human visitation and are managed in a way that may contribute to local economies through promoting educational and recreational tourism while promoting conservation efforts. The surrounding areas of a National Park may be for consumptive or non-consumptive use, but should nevertheless act as a barrier for the defence of the protected area's native species and communities to enable them to sustain themselves in the long term.
Category III – Natural Monument or Feature		Represents comparatively smaller areas that are specifically allocated to protect a natural monument and its surrounding habitats, including natural geological or geomorphological features; culturally influenced natural features; natural cultural sites; or cultural sites with associated ecology.

Category IV – Habitat/Species Management Area	Like Category III, this focuses on more specific areas of conservation but in relation to an identifiable species or habitat that requires continuous protection rather than that of a natural feature. These protected areas will be sufficiently controlled to ensure the maintenance, conservation and restoration of particular species and habitats – possibly through traditional means – and public education about such areas is widely encouraged as part of the management objectives. Habitat or species management areas may also be used in the management of fisheries, and exist as a fraction of a wider ecosystem or protected area and may require varying levels of active protection. Management measures may include the prevention of poaching, creation of artificial habitats, halting natural succession and supplementary feeding practices.
Category V – Protected Landscape/Seascape	A protected area which covers entire bodies of land or ocean with a more explicit management plan in the interest of nature conservation, but is more likely to include a range of for-profit activities. The main objective is to safeguard regions that have built up a 'distinct character' in regard to their ecological, biological, cultural or scenic value. In contrast with previous categories, Category V allows a higher level of interaction with surrounding communities who are able to contribute to the area's management and engage with the natural and cultural heritage it embodies through a sustainable outlook. As a result, protected landscapes and seascapes may be able to accommodate contemporary developments such as ecotourism at the same time as maintaining the historical management practices that may procure the sustainability of agro-biodiversity and aquatic biodiversity.
Category VI – Protected Area with sustainable use of natural resources	A generally more encompassing classification that is focused on the mutually beneficial correlation between nature conservation and sustainable management of natural resources, taking into account the livelihoods of those who are dependent on both. A wide range of socioeconomic factors are taken into consideration in creating local, regional and national approaches to using natural resources as a tactic to assist sustainable development rather than hinder it.

This categorisation method is recognised on a global scale by national governments and international bodies such as the United Nations and the Convention on Biological Diversity (CBD).

Did you know?

Protected areas will usually encompass several other zones that have been deemed important for particular conservation uses, and the range of natural resources that any one protected area may guard is vast. The main aim of declaring protected areas is to conserve species or the relationships between species. They are, however, similarly important for conserving sites of cultural or indigenous importance and considerable reserves of natural resources, such as carbon stocks, rainforest and mountains, and species such as birds, cetaceans and fish.

Key points

- Protected areas are a management tool used to effect in-situ conservation of the Earth's environment, ecosystems and species.
- Protected areas range in their types – terrestrial, marine or transboundary – as well as their objectives.
- The most commonly used system of defining protected areas, as well as categorising their objectives, is the IUCN system.

∞ Links

See 3.20 on in-situ conservation, 3.27 on community-based management, and 3.30 to 3.32 on environmental and conservation law.

See also the following section on marine and transboundary protected areas.

On completion of this section, you should be able to:

- Identify and explain the role of marine protected areas

- Identify and explain the role of transboundary protected areas.

Marine protected areas (MPAs)

Marine protected areas are included in the general IUCN for protected areas. An MPA may be either a totally marine area with no significant terrestrial parts; an area containing both marine and terrestrial components, or a marine ecosystem that contains land and intertidal (land that is frequently covered by water) components only. For example, a mangrove forest would contain no open sea or ocean marine environment, but its river-like marine ecosystem nevertheless matches the definition. Typical restrictions in MPAs include ones on fishing, oil and gas mining and tourism. Other restrictions may limit the use of ultrasonic devices like sonar (which may confuse the guidance system of cetaceans), development and construction. Some fishing restrictions include so-called 'no-take' zones, which means that no fishing is allowed. In terms of categories, the following apply:

Category		Description
1	Category Ia – Strict Nature Reserve	A marine reserve usually connotes 'maximum protection', where all removals of resources are strictly prohibited. In some countries such as Belize, marine reserves allow for low-risk forms of exploitation to sustain local communities. Though this suggests an area of lower protection, these reserves maintain their categorical status which iterates their biological importance.
	Category II – National Park	Marine parks lay a high emphasis on the protection of ecosystems but have been known to allow light human use. A marine park may prohibit fishing or extraction of resources of any kind, but could possibly allow recreation. However some marine parks, such as the Folkstone Marine Park (Barbados), Soufriere Marine Management Area (SMMA) (St Lucia) and Soufriere-Scotts Head Marine Reserve (Dominica) are zoned and activities such as fishing are only allowed in particularly low-risk areas.
	Category III – Natural Monument or Feature	Established to protect historical sites such as shipwrecks and cultural sites such as indigenous fishing grounds.
	Category IV – Habitat/ Species Management Area	Established to protect a certain species, to benefit fisheries, rare habitat, as nursing grounds for fish, or to protect entire ecosystems.
	Category V – Protected Landscape/Seascape	Limited active management assigned, as with protected landscapes.

These categories may also encompass World Heritage Sites (WHS) – areas exhibiting extensive natural or cultural history, and Ramsar sites which meet the definition of 'wetlands' under the 1971 Ramsar Convention. There are approximately 500 MPAs in the Caribbean region (Guarderas *et al.*, 2008), with coral reefs being the best represented ecosystem.

The Caribbean Challenge Initiative (CCI) is the first conservation initiative embraced by governments across the region. The aim is to have at least 20 per cent of their near-shore marine and coastal environments in national marine protected areas systems by 2020 and to create new sustainable finance mechanisms (such as tourism fees), dedicated solely to funding park management.

Transboundary protected areas

A transboundary protected area (TBPA) is a protected area that extends across boundaries of multiple countries, and where the political border sections that are enclosed within its area are largely ignored. An example of a transboundary protected area in the Caribbean region is the Mesoamerican Barrier Reef System (MBRS). The reef system extends a marine region that stretches over 1000 km along the coast of four countries: Mexico, Belize, Guatemala and Honduras, and includes various protected areas and parks including the Belize Barrier Reef, which is home to approximately 80 per cent of the MBRS.

Did you know?

The Belize Barrier Reef is the largest barrier reef in the northern hemisphere and the second largest barrier reef in the world, and along with Belize's three offshore atolls, several hundred sand cays, mangrove forests, coastal lagoons and estuaries is collectively termed the Belize Barrier Reef Reserve System. It was designated a World Heritage Site by UNESCO in 1996.

 Links

See the previous section on protected area systems.

Key points

- An MPA may be either a totally marine area with no significant terrestrial parts; an area containing both marine and terrestrial components, or a marine ecosystem that contains land and intertidal (land that is frequently covered by water) components only.

- There are approximately 500 marine protected areas (MPAs) in the Caribbean region (Guarderas *et al.*, 2008/2011).

- A transboundary protected area (TBPA) is a protected area that extends across boundaries of multiple countries, and where the political border sections that are enclosed within its area are largely ignored.

3.27 Community-based natural resource management

Learning outcomes

On completion of this section, you should be able to:

- identify the purpose of community-based natural resource management in the Caribbean region

- outline the role of community-based natural resource management as a tool for natural resource management and conservation in the Caribbean region.

Community-based management is an approach to natural resource management that takes into account the human communities living around or within a resource. It has become increasingly more common since the 1980s, and generally focuses on interdependent ecological, economic and social elements, and has therefore been categorised as one form of sustainable development.

The role of the community in natural resource management

Throughout the islands of the Caribbean, initiatives are underway to engage communities in co-management of natural resources. The stated rationale is often that community involvement can help to reduce the degradation of marine and terrestrial biodiversity, address resource use conflicts, improve the community's quality of life and provide opportunities for economic activity. Other goals include improved governance through building stronger community institutions and increased community capacity, empowerment and voice, which can in turn provide a vehicle for strengthening local governance in other spheres of social and economic development (Borrini-Feyerabend *et al.*, 2004).

Case studies in Community-based natural resource management in the Caribbean

Natural resource	Comments
Fisheries	- Community-based management of fisheries, mangrove areas and marine reserves has also been initiated in St Lucia, in the Soufriere Marine Management Area (SMMA) which involves fishing, tourism and other interests in a series of zoned fishing priority and marine reserve areas. - In Dominica, the Soufriere/Scotts Head Marine Reserve (SSMR) is growing in popularity as a tourism attraction and no-take marine reserve in addition to having a long-established artisanal fishery. It has been placed under a local area management authority (LAMA) following representation by a wide range of interests including fisheries, tourism and environmentalists. - In St Vincent & the Grenadines the emphasis is on safeguarding traditional management methods such as beach seining operations by means of fisherfolk organisations, mostly cooperatives. The expected impacts of community-based management initiatives include improved quality of fishery management decisions and their implementation; empowerment and enhanced capability of communities to manage fishery resources for sustainability; and increased contribution of the fisheries sector to economic development.
Watershed management	- The Jamaica Conservation and Development Trust co-manages the John Crow and Blue Mountain National Park. One of its goals is to increase support and improve natural resource management in the park's buffer zone communities, leading to ecosystem conservation and poverty alleviation. This includes working with buffer zone communities to plan and implement projects with both conservation and livelihood components, such as reforestation, community tourism and sustainable agriculture (Beale, 2010). - The Saint Lucia Forestry Department conceived of local WaterCatchment Groups as a mechanism to involve communities in watershed management, but it has not been able to

	secure consistent funding to support the Groups. Yet research has found early evidence of positive contributions to water quality, water quantity and community awareness and concluded that investing a small percentage of water revenues in further organisational capacity building could generate sustained and enhanced results (Pantin, Reid and Maurice, 2006). ▪ 20 years after their creation under an initiative of the Wildlife Division of the Forestry Department in Trinidad and Tobago, Nature Seekers and the Grande Riviere Tourism Development Company have become world famous for their turtle research and protection programmes and are now important local employers, generating revenue from a range of ecotourism, agricultural and forest management initiatives (Trewenack, 2010).
Forest management	▪ Jamaica's forest legislation and policy makes specific provisions for stakeholder participation in management and decision making about the use of forest resources. Stakeholder and community participation is identified as key to the implementation strategies towards meeting the country's forestry and watershed management goals and placing participatory forest management in a livelihoods framework. The Local Forest Management Committees (LFMCs) are the main vehicle for doing so at the local level. Their duties include monitoring and assisting with management; public education and mobilisation; and making regulations and proposing incentives for local conservation practices (Brown and Bennett, 2010).
Tourism	▪ As part of the National Reforestation and Watershed Rehabilitation project in Trinidad and Tobago community groups used their existing conservation and ecotourism experience to insist on planting native species, inter-cropping with fruit trees, and developing trails, enhancing both the conservation and the livelihoods outcomes (McDermott, 2010; Trewenack, 2010). ▪ NGOs such as Grenada Community Development Agency (GRENCODA) and Agency for Rural Transformation (ART) have worked alongside government agencies to build the capacity of community-based organisations, helping in the development of a proposed community tourism policy and implementing community tourism projects, mainly based on the use of natural resources, such as trails, fish festival, craft, beach management, etc. (Government of Grenada, 2006).

The Soufriere Marine Management Area (SMMA) is also a World Heritage Site of natural significance. The stunning backdrop of the Pitons, and the importance of the marine and coastal area to the populace of Soufriere, were some of the reasons leading to the designation of the area in 2004. In 2012, the area narrowly missed being placed on UNESCO's 'world heritage in danger' list.

∞ Links

See 3.25 on protected area management categories and objectives, and 3.29 on public participation and advocacy.

Key points

- Community-based management has evolved as a valuable tool in natural resource management in the Caribbean region.

- It has expanded the model of natural resource management from a largely top-down approach, to one which is more participatory and inclusive in nature.

- Successful examples of community-based management may be found in many sectors of natural resource management, including fisheries, watershed management, forestry, protected area management and tourism.

Learning outcomes

On completion of this section, you should be able to:

- identify the economic instruments which may be used for natural resource management and conservation in the Caribbean region

- explain the use of these economic instruments.

Did you know?

An example of a user fee is the charge to enter Dunn's River Falls in Ocho Rios, Jamaica or Harrison's Cave in Barbados, where the fees and charges to users are used to partially offset the costs of management of these public facilities, and supplement funds provided by the respective governments.

Performance bonds

The general principle of performance bonds is that the supervising government agency is guaranteed sufficient funds, in the form of a bond or security, to cover the cost of rehabilitation in the event of failure by the enterprise concerned.

Economic instruments

Economic instruments for environmental and natural resource management can be defined as administrative mechanisms adopted by government agencies to influence the behaviour of people who value the natural environment, make use of it, or cause adverse impacts as a side-effect of their activities. Economic instruments are one class of environmental policy tool – which may be contrasted with other tools such as regulatory and technological instruments. Regulatory instruments are in the form of laws and mechanisms which specify physical standards to be attained, while technological instruments specify technology to be adopted. A broad distinction can be drawn between these two categories of direct regulations – commonly described as command-and-control mechanisms – and economic instruments. Command-and-control mechanisms are based primarily on legislative and regulatory provisions and are implemented through directives from regulatory authorities. Economic instruments, on the other hand, set charges for environmental use of degradation, but do not specify the environmental quality to be obtained or the equipment to be used.

Natural resources and the environment are considered by economists to be in limited supply, and their over-use and consequent resource degradation are seen to be symptomatic of 'market failure'. This is because users of natural resources and the environment are not held responsible for the full costs of resource use, primarily because of the 'public goods' characteristics of the environment, which make it difficult to control access or exclusivity of use. General areas of application of economic instruments include pollution control (discharge of solids, effluents and gases); noise and thermal pollution; physically renewable resources (water, renewable energy); non-renewable resources (minerals and other materials); semi-renewable resources (agricultural land, groundwater supplies); biologically renewable resources (forestry, fisheries); conservation reserves and natural areas (national parks and wilderness areas); areas of aesthetic or heritage value and biodiversity and natural ecosystems.

User fees

User fees are applied by state and private operators for the use of natural environmental amenities for recreation, scientific research and education. Areas for which fees may be charged include national parks, recreation areas and conservation reserves. In principle, fees could be used to ration use of such resources, especially to reduce congestion and resource degradation. However, this is rarely the case in practice, and in most cases fees are imposed to help cover management costs.

Environmental levies and taxes

Environment taxes consist of a special levy to finance environmental improvement programmes and projects. A potentially efficient instance of such a tax, unrelated to incentive effects, is a levy designed as a front-end capital financing measure. An example of such a levy was that imposed in Barbados to defray the cost of the disposal of refuse generated by the use of goods imported into Barbados. An environmental levy was charged at the rate of one per cent of the customs value on most goods, with a very limited number of goods carrying a specific rate of environmental levy, e.g. motor vehicles. However, this levy was repealed in 2011. A tax exemption to create significant economic incentives for paper manufacturers to use recycled paper is an example of an incentive which may lead to major investments in manufacturing facilities specially designed to handle recycled paper inputs.

Environmental incentives for biodiversity conservation

Impacts on biodiversity are pervasive and cross many jurisdictional boundaries, and so management controls must therefore be comprehensive. The declaration of conservation reserves is only one of many options for managing biodiversity, but this inevitably means introducing incentives and controls in the private as well as the public sector. The use of economic incentives to protect and encourage biodiversity is a comparatively recent focus of policy. Such incentives are usually implemented in conjunction with a wide range of regulatory and other measures.

Economic instruments may be used to manage populations of species harvested commercially in connection with fisheries management. The use of individual transferable quotas and controls over harvesting are relevant examples. For invasive species which can threaten ecosystems or species, the allocation of licences to professional cullers of natural populations is also a possible use. Thus, for example, the bounty system put in place by the Ministry of Agriculture of Barbados grants B$0.50 per pound of giant African snail collected by the public. Donations may also be sought from the general public, through fundraising for specific causes. For example, tourists may make direct personal contributions at particular sites. Donations may be sought through animal sponsorship schemes, under which a person 'adopts' a particular animal in a species conservation or rehabilitation programme. Certificates and other tokens can be used to reinforce the identification. Private corporations wishing to demonstrate their environmental responsibility frequently provide direct financial support for the protection of natural species and habitats.

Links

See 3.8 for more information on economic instruments.

Key points

- Economic instruments operate through market processes or other financial incentives.

- Each class of environmental or natural resource management problem has its own special properties or attributes, thus the suitability and kind of economic instrument must be considered on a case-by-case basis.

- Economic instruments may be used in protected area management, in fisheries, to encourage recycling, for biodiversity conservation and to promote restoration and rehabilitation schemes.

CASE STUDY

Economic instruments for the management of waste in the Caribbean region

Economic instruments are an increasingly common feature of waste management in the Caribbean region. They are being used in an attempt to reflect the total environmental cost of the management and disposal of waste. First, governments may recover administrative costs by charging licence fees to industries which generate significant quantities of waste – such as hotels, factories and breweries. This was done in Barbados, under the environmental levy system that was discontinued in 2011.

Several categories of instruments have been used in this regard, and waste management programmes also feature the introduction of cost recovery measures for services provided by the waste management agency. States such as Barbados and the British Virgin Islands also have mechanisms whereby deposit refunds on recyclable containers

seek to address the problems of environmental degradation caused by improper disposal of containers to the environment. The introduction of modest payments by manufacturers in many Caribbean states for recycled cans and bottles has resulted in improved collection services.

Finally, most Caribbean countries have enacted an environmental levy act. There are two main purposes behind this instrument – firstly, the environmental levy is to address issues associated with cruise ship tourism (pollution from the ships, littering and an increase in consumers), since the passengers on cruise ship use a countries' facilities, but there is often no guarantee of a financial investment by these visitors. Second, the environmental levy may be imposed for the importation of articles such as plastics, computer waste etc., destined for disposal in overcrowded landfills – for example as stipulated in the 1999 Environmental Protection Levy Act of St Lucia.

3.29 The use of education, awareness, advocacy and training

Did you know?

The 1977 Tbilisi Declaration was the product of the Intergovernmental Conference on Environmental Education in Tbilisi, Georgia. The Declaration emphasised the role of environmental education in preserving and improving the global environment and sought to provide the framework and guidelines for environmental education.

Education, public awareness, advocacy and training

The roots of environmental education can be traced back as early as the 18th century when Jean-Jacques Rousseau stressed the importance of an education that focuses on the environment in his work *Emile: or, On Education*. Several decades later, Louis Agassiz, a Swiss-born naturalist, echoed Rousseau's philosophy as he encouraged students to 'study nature, not books'. These two influential scholars helped lay the foundation for a concrete environmental education programme, known as nature study, which took place in the late 19th century and early 20th century. Ultimately, the first Earth Day on 22 April 1970 – a national teach-in about environmental problems, which is still widely celebrated – paved the way for the modern environmental education movement.

Internationally, environmental education gained recognition at the UN Conference on the Human Environment held in Stockholm, Sweden, in 1972, when Principle 19 of the Stockholm Declaration underscored the importance of environmental education as a tool to address global environmental problems. Supplemented by the 1979 Belgrade Charter, the goals, objectives, characteristics, and guiding principles of environmental education would best be summarised in the 1977 Tbilisi Declaration, which sought to provide the framework and guidelines for environmental education. The Tbilisi Conference laid out the role, objectives and characteristics of environmental education, and provided several goals and principles for environmental education. In September 2012, in honour of the 35th anniversary of the Tbilisi Declaration, the Intergovernmental Conference on Environmental Education for Sustainable Development released the Tbilisi Communiqué – 'Educate Today for a Sustainable Future'.

Environmental education refers to the act of teaching about the features, benefits and functions of the environment and its components, and the relationship between humans and this environment. The ultimate goal is for humans to appreciate the complex and inextricable links between themselves and the environment, and make prudent decisions aimed at sustainable use and development of natural resources. Environmental education may take the form of formal education via the primary, secondary or tertiary curricula, or by less formal means – using various types of media to promote awareness on general or specific environmental issues. Awareness efforts may utilise well orchestrated events, such as declaring 2010 the Year of Biodiversity, public service announcements, use of social networks, billboards and flyers, and incentive measures such as competitions. Environmental education aims to:

■ promote awareness and sensitivity about the environment and environmental challenges;

■ help people develop a knowledge and understanding of the environment and their place in it;

■ produce an awareness of the need to maintain environmental quality;

■ provide skills to help in mitigating environmental problems;

■ encourage participation in environmental issues and environment-related programmes.

One of the key methods that has been used to extend education on environmental awareness is by training a group of individuals on the background knowledge, skills and practical experience of specific environmental issues, and then using this core group to propagate the experience to other groups in society. For example, in 2007 the UNEP-CAR/Regional Coordination Unit launched and coordinated a 'Training of the Trainers' for managers of marine protected areas (MPAs), through which managers are not only trained in all aspects of MPA management, but also adult education techniques to conduct local and tailored training activities in their respective MPAs (UNEP/CEP, 2013).

Environmental education has evolved into an environmental movement which has a broader focus than merely imparting awareness on the environment. It includes environmental activism, environmental advocacy, and environmental justice, and can be championed by diverse scientific, social and political actors. Environmental advocacy involves working to influence public policy in social, economic, political and cultural spheres in order to bring about justice and positive change on environmental issues. Advocates seek to protect the public from environmental hazards and to protect features of the natural environment. Advocates often organise in groups to champion a cause and work to implement changes which will have a positive effect on the environment. Examples of advocacy groups include Greenpeace and Friends of the Earth. Notable examples in the Caribbean are the Belize Alliance of Conservation Non-Governmental Organisations (Belize), Fishermen and Friends of the Sea (Trinidad & Tobago), the Save Guana Cay Reef Association (the Bahamas), the Virgin Islands Environmental Council (the Virgin Islands), and the Guyana Marine Turtle Conservation Society (Guyana), to name a few. Many of these groups have been involved in legal action, as well as lobbying governments and educating the public on specific issues of environmental concern. The hunger strike staged by environmental activist and UWI St Augustine lecturer Dr Wayne Kublalsingh from 16 November to 5 December 2012 in front of the Office of the Prime Minister of Trinidad and Tobago is a recent example of environmental advocacy in the region. The strike was orchestrated over the construction of a highway in south Trinidad on certain conditions, and ended only after reworked terms of reference had been accepted for the works on the Debe to Mon Desir section of the highway.

Key points

- Environmental education, awareness and advocacy has long been recognised as a critical ingredient in natural resource management.

- It is central to several aspects of contemporary natural resource management, including the environmental impact assessment (EIA) and community-based natural resource management.

- Although environmental education is an aspect of natural resource management that has traditionally been neglected in the Caribbean region, it has recently become more common and widespread.

 Links

Education, advocacy and awareness is a key ingredient in community-based participation in natural resource management, as well as in the EIA process.

On completion of this section, you should be able to:

- identify key international agreements concerned with the management and conservation of natural resources

- understand the use and application of international agreements as a tool for the management and conservation of natural resources

- explain how environmental law derived from international law is implemented and enforced by Caribbean states by means of policies and legislation.

Did you know?

There are over 900 international and regional agreements concerning the environment!

∞ Links

The obligations under these conservation agreements lead to many in-situ and ex-situ conservation practices. See 3.20 for more on these types of conservation practices.

Did you know?

At the 1992 Rio Conference, delegates agreed to and signed two of the principal international agreements on the environment. To date, the United Nations Convention on Biological Diversity and the United Nations Framework Convention on Climate Change enjoy almost universal acceptance in the Caribbean region.

Environmental and conservation agreements and laws: From international and regional to national implementation

In the aftermath of the 1972 Stockholm Conference the United Nations Environmental Programme (UNEP) began to produce legally binding agreements concerned with addressing a variety of environmental problems. These came to be known as multilateral environmental agreements (MEAs) – legally binding treaties between three or more states relating to environmental issues such as species loss, habitat degradation and other activities by humans on the environment. Agreements may be called bilateral environmental agreements (BEAs) if the agreement is between two nation states.

When nations get involved in MEAs, they usually pass laws to place these agreements into law. In addition, they may pass laws for subjects such as public health, land use planning and coastal zone management, aimed at reducing the impacts of human activity in two main areas: natural resource and species conservation, and pollution control. Effective enforcement – the set of actions that the government can take to promote compliance with environmental law– is key to ensuring that the goals of environmental statutes are realised.

International and regional law also informs a nation's environmental policy where a state devises strategies aimed at addressing problems arising from the human interaction with the environment. Examples of policies developed by Caribbean states for environmental protection and the use of natural resources include National Environmental Action Plans (Anguilla, 2005), Integrated Coastal Zone Management (ICZM) Plans (Barbados, 1996), Beach Policy (Jamaica, 2000) National Mangrove Management Plans (Guyana, 2010) and National Forest Policy (Jamaica, 2001; 2012). Brief concept papers – for example on renewable energy are also another form of policy, which may provide general vision for future developments.

A policy document which is common to all Caribbean states is the National Biodiversity Strategy and Action Plan (NBSAP), the principal instruments for implementing the 1992 Convention on Biological Diversity at the national level (Article 6). The Convention requires countries to prepare a national biodiversity strategy (or equivalent instrument) and to ensure that this strategy is mainstreamed into the planning and activities of all those sectors whose activities can have an impact on biodiversity.

International Conservation Agreements

The 1992 United Nations Convention on Biological Diversity

The 1992 Convention on Biological Diversity is an MEA which lists in its preamble three main goals: the conservation of biological diversity (or biodiversity); the sustainable use of its components; and the fair and equitable sharing of benefits arising from genetic resources. The Convention is a landmark document in international law on the environment, since it recognised for the first time in international law that the conservation of biological diversity is an integral part of the sustainable development process, and seeks to integrate the three pillars of sustainable development. The agreement is extensive and addresses all ecosystems, species, and genetic resources, and their conservation by in-situ and ex-situ means. Areas covered by the Convention include the management of invasive species, environmental education and awareness and environmental impact assessments. Caribbean states may therefore utilise the agreement to manage issues of natural resource use, conservation, ecosystem degradation and species extinction.

Two protocols (or complementary agreements) complete the Convention: the 2010 Nagoya Protocol which sets principles for the fair and equitable sharing of the benefits arising from the use of genetic resources, notably those destined for commercial use, and the 2003 Cartagena Protocol on Biosafety, which addresses the rapidly expanding field of biotechnology and biosafety issues.

The 1971 Ramsar Convention

The Convention on Wetlands of International Importance, especially as Waterfowl Habitat, commonly known as the Ramsar Convention after the name of the city where the treaty was signed in 1971, is an MEA addressing the conservation and sustainable utilisation of wetlands. It is therefore a crucial agreement for the conservation and management of wetland and mangrove ecosystems in the Caribbean region – and by extension, the myriad species which they host.

Wetlands are defined to include areas of water – fresh, brackish or salt, including areas of marine water the depth of which at low tide does not exceed 6 metres. It also includes fish ponds, rice paddies and salt pans. There are several Ramsar sites in the Caribbean, including the Caroni and Nariva Swamps in Trinidad; for example the Graeme Hall Swamp in Barbados.

MARPOL 73/74

The International Convention for the Prevention of Pollution From Ships, 1973 (MARPOL), as modified by the Protocol of 1978, is an international treaty which was designed to minimise pollution of the seas, including dumping, oil and exhaust pollution. The Convention is one of the most important international marine environmental conventions, and most Caribbean States are parties to the Convention. This is of vital importance to the Caribbean Sea and coasts and marine ecosystems.

The stated object of the agreement is to preserve the marine environment through the complete elimination of pollution by oil and other harmful substances and the minimisation of accidental discharge of such substances. Accordingly, the agreement contains six annexes, concerned with preventing different forms of marine pollution from ships: Annex I (oil); Annex II (noxious liquid substances carried in bulk); Annex III (harmful substances carried in packaged form); Annex IV (sewage); Annex V (garbage) and Annex VI (air pollution). A state that becomes party to MARPOL must accept Annex I and II, but Annexes III–VI are voluntary annexes.

Did you know?

There is global concern over the loss of wetlands and their associated ecological services and biodiversity because of the progressive encroachment on and loss of wetlands to urbanisation, aquaculture and habitat conversion. The main tenet underpinning the Ramsar Convention is therefore the concept of 'wise use', which lies at the heart of the MEA, and which aims to balance conservation and the sustainable use of wetlands and their resources.

Key points

- As a result of the 1972 Stockholm and 1992 Rio Conferences, international efforts were put in place to create instruments to address various threats affecting the Earth's environment, ecosystems and species.

- One of the key methods is the use of multilateral environmental agreements (MEAs), which are contracts between more than two states on a specific environmental issue.

- MEAs have had a tremendous impact on the law of individual states in the Caribbean region.

Did you know?

CITES is one of the largest and oldest conservation and sustainable use agreements in existence!

Addressing specific threats to species and ecosystems

Environmental threats are not experienced only at ecosystem level; individual species are often the first affected when degradation of ecosystems takes place. Additionally, in the last 20 years, focus has shifted to the effects of human-caused or anthropogenic effects on climate change, and its resultant effects such as desertification. Specific agreements have therefore been drawn up to address these issues, and they are discussed below.

International conservation agreements

The Convention on International Trade in Endangered Species of Wild Fauna and Flora (CITES) is a multilateral treaty, drafted as a result of a resolution adopted in 1963 at a meeting of members of the International Union for Conservation of Nature (IUCN). The convention was opened for signature in 1973, and entered into force on 1 July 1975. Its aim is to ensure that international trade in specimens of wild animals and plants does not threaten the survival of the species in the wild, and it accords varying degrees of protection to more than 34,000 species of animals and plants. The 1992 United Nations Framework Convention on Climate Change arose out of the 1992 Rio Conference, and, along with its main amendment the Kyoto Protocol, addresses the sources and effects of the environmental threat of global warming and climate change. One such effect – desertification – is addressed by the 1994 United Nations Convention to Combat Desertification.

The 1973 Convention on International Trade in Endangered Species of Wild Fauna and Flora

CITES works by subjecting international trade in specimens of selected species to certain controls. All import, export and re-export covered by the Convention has to be authorised through a licensing system. The principal mechanism is the listing of species into three Appendices. With the coming into existence of the Convention on Biological Diversity (CBD) in 1992, there has been increased collaboration between the two MEAs.

Appendix	Explanation	Examples in the Caribbean
Appendix I	Species threatened with extinction. Commercial trade in wild-caught specimens of these species is prohibited, permitted only in exceptional licensed circumstances (e.g. for ex-situ conservation measures).	<ul><li>all species of marine turtles</li><li>manatees (Sirenia)</li><li>the jaguar (*Panthera onca*) (Belize, Suriname and Guyana).</li></ul>
Appendix II	Contains species not necessarily threatened with extinction, but that may become so unless trade in specimens is regulated. International trade in Appendix II species may be authorised by the granting of an export permit or re-export certificate.	<ul><li>the green iguana (*Iguana iguana*) (Jamaica)</li><li>the queen conch (*Strombus gigas*) (Belize, Jamaica and most Caribbean states)</li><li>bigleaf mahogany (*Swietenia macrophylla*) (Belize).</li></ul>

Appendix III	Species that are listed after one member country has asked other CITES Parties for assistance in controlling trade in a species. The species are not necessarily threatened with extinction globally.	■ none to date within the Caribbean region ■ however, the Central American state of Costa Rica has requested the listing of the two-toed sloth (*Choloepus hoffmanni*) and the alligator snapping turtle (*Macrochelys temminckii*) by the United States.
Split listing	Species may be split-listed, meaning that some populations of a species are on one Appendix, while some are on another.	■ the St Vincent parrot (*Amazona guildingii*), endemic to St Vincent and the Grenadines, is listed on Appendix I and II of CITES ■ the African elephant (*Loxodonta africana*) is currently split-listed, with all populations except those of Botswana, Namibia, South Africa and Zimbabwe listed in Appendix I. Those of Botswana, Namibia, South Africa and Zimbabwe are listed in Appendix II.

The 1992 United Nations Framework Convention on Climate Change (UNFCCC) and the 1997 Kyoto Protocol

According to Article 2 of the UNFCCC, the objective of this treaty is to 'stabilise greenhouse gas concentrations in the atmosphere at a level that would prevent dangerous anthropogenic interference with the climate system'. The agreement has become increasingly important in light of concerns about global climate change. The main mechanism of the treaty is to divide parties to the agreement into two broad categories – Annex I which are primarily industrialised (developed) countries and 'economies in transition' states, and Annex II, which are mostly low-income developing states. Caribbean countries fall into this Annex. The most important protocol of the Convention is the 1997 Kyoto Protocol, which was adopted by the parties to the UNFCCC in 1997, and entered into force in 2005. As part of the Kyoto Protocol, many developed countries have agreed to legally binding limitations/reductions in their emissions of greenhouse gases in two commitment periods. The first commitment period applies to emissions between 2008 and 2012, and the second commitment period applies to emissions between 2013 and 2020.

The 1994 United Nations Convention to Combat Desertification (UNCCD)

The aim of the 1994 'United Nations Convention to Combat Desertification in Those Countries Experiencing Serious Drought and/or Desertification, particularly in Africa' (UNCCD) is to combat desertification and mitigate the effects of drought on the world's countries, people and ecosystems. Many Caribbean states have land use issues arising from natural disasters such as the combined effects of Hurricanes Ivan and Emily on Grenada, as well as poor land use practices such as beach mining and unsustainable agriculture. They have therefore implemented sustainable land management programmes which exemplify the tenets of the agreement.

Key points

- In recent years, focus has shifted to the effects of anthropogenic-driven effects on climate change, and its resultant effects.
- Agreements such as UNFCCC and UNCCD have come about as a result.

Did you know?

The 1973 Convention on International Trade in Endangered Species of Wild Fauna and Flora (CITES) is one of the largest and oldest conservation and sustainable use agreements in existence. The agreement seeks to regulate the international trade in specimens of wild animals and plants by according varying degrees of protection to more than 34,000 species of animals and plants. Many wild and important commercial species in the Caribbean are now included on the CITES list.

∞ Links

The obligations under these conservation agreements lead to many in-situ and ex-situ conservation practices, such as protected areas, land use management and seed banks. See 3.20 for more on these conservation practices.

Did you know?

The Caribbean Sea is categorised as a semi-enclosed sea!

Did you know?

Regional agreements on the environment address issues which are of particular importance to a specific region or place. The Caribbean region has one principal agreement with three supporting protocols which addresses the management and conservation of marine, coastal and inland ecosystems, as well as species in the region.

Regional conservation agreements

The 1983 Cartagena Convention

The 1983 Convention for the Protection and Development of the Marine Environment of the Wider Caribbean Region is a comprehensive framework agreement for the conservation and management of the marine environment in the Wider Caribbean Region. The treaty focuses specifically on those waters making up the Caribbean Sea, the Gulf of Mexico and parts of the Atlantic Ocean, referred to as the Convention Area. It provides the legal framework for cooperative regional and national actions by Caribbean states. Most Caribbean states are parties to the agreement, and have implemented its policies and actions in their laws.

Cartagena has three Protocols – dealing with oil spills, protected areas, and land-based pollution –and is the key regional instrument by which the Caribbean manages its sea, inland waters, the services which they provide, its species and man's use of these resources.

In addition to general obligations and institutional arrangements regarding the management of the Convention area, the Cartagena Convention lists the sources of pollution which have been identified by Caribbean states to require regional and national action for their control. The Convention also identifies environmental management issues for which cooperative efforts are necessary – specially protected areas and wildlife, cooperation in cases of emergency, environmental impact assessment and scientific and technical cooperation.

Did you know?

This multilateral environmental agreement was not intended as a stand-alone agreement, or designed to supersede existing agreements, but is partly designed to supplement other multilateral treaties in place, such as MARPOL 73/78, Ramsar, and the Convention on Biological Diversity (CBD), and to complement the 1982 Montego Bay Convention on the law of the sea.

The SPAW Protocol

Article 10 of the Cartagena Convention provides for conservation and management of marine protected areas. The Article obliges its parties to take measures to protect and preserve rare or fragile ecosystems, as well as the habitat of depleted, threatened or endangered species, in the Caribbean region. The SPAW Protocol incorporates several innovative ideas which are central to sustainable development, including the precautionary principle, and explicit recognition of links between habitat protection (particularly marine protected areas) and the protection of endangered species.

In establishing protected areas under the Protocol, parties are expected to conserve, maintain and restore the natural environment, which includes coastal and marine ecosystems, habitats and their associated ecosystems. These features are critical to the survival and recovery of endangered, threatened or endemic species of flora or fauna, and the productivity of ecosystems in the Caribbean region.

In a similar manner to CITES, Article 11 of the Protocol establishes three Annexes. However, unlike CITES, SPAW differentiates between flora and fauna whose use is strictly prohibited, and then addresses those species to be regulated in the third Annex.

Annex	Explanation	Examples
Annex I	Includes threatened or endangered plant species for which any form of destruction or disruption (picking, gathering, uprooting, cutting, possession, trade, etc.) must be banned in order to guarantee their protection and if need be their recovery.	■ morning glory (*Ipomoea walpersiana*) ■ rosewood (*Ottoschulzia rhodoxylon*) ■ Jost Van Dyke's Indian mallow (*Abutilon virginianum*).
Annex II	Lists threatened or endangered animal species for which, again, any form of destruction or disruption (capture, possession, killing, trade, etc. must be banned for their protection and recovery).	■ all species of marine turtles ■ all species of manatees (Sirenia) ■ all species of whales and dolphins ■ Jamaican ground iguana (*Cyclura collie*) ■ Oncilla, tiger cat (*Felis tigrina*).
Annex III	A list of animal *and* plant species for which special measures must be taken to ensure their protection and recovery whilst authorising and regulating the use of these species.	■ all species of hard and soft corals ■ red, black and white mangroves ■ several species of seagrass ■ queen conch (*Strombus gigas*) ■ scarlet ibis (*Eudocimus ruber*) ■ prickly pear (*Opuntia macracantha*).

Recent proposals for listing in Annex III have included commercially valuable species such as the Nassau grouper (*Epinephelus striatus*) and the spiny lobster (*Panulinus argus*).

The Oil Spills and LBS Protocols

In addition to the establishment of marine and coastal protected areas, Articles 5, 6 and 8 of the Convention include measures which oblige parties to take all practical measures to prevent, reduce and control pollution from ships, by dumping, and from seabed activities. Accordingly, the Oil Spills Protocol addressed the another source of concern for Caribbean Sea in the form of the marked increases in oil, natural gas and shipping activities, all of which pose major challenges to the effective management of the marine resource in a sustainable manner. The Protocol applies to oil spill incidents which have resulted in, or which pose a significant threat of, pollution to the marine and coastal environment of the Wider Caribbean Region, and also pollution incidents which adversely affect the related interests of one or more of the contracting parties to the agreement.

Finally, it is widely recognised that the largest threat to the coastal and marine zone is land-based sources of pollutants. The threat from land-based sources has been identified under the Cartagena Convention, in the third and most recent of its protocols – the Protocol Concerning Pollution from Land-Based Sources, known as the LBS Protocol. This Protocol obliges each party to take appropriate measures to prevent, reduce and control pollution of the Convention Area from land-based sources and activities. In achieving this goal parties are to use the best available means at their disposal. Parties must also use the most appropriate technology and management approaches, such as integrated coastal management. The LBS Protocol is also closely linked with the SPAW Protocol through the definition of 'Class I waters'. Class I waters are classified as those that are particularly sensitive to the impacts of domestic wastewater, and include all those that provide habitat for species protected under the SPAW Protocol, protected areas and waters used for recreation.

∞ Links

The obligations under these conservation agreements lead to many in-situ and ex-situ conservation practices, such as protected areas, land use management and seed banks. See 3.20 for more information on these conservation practices.

Key points

- In the Caribbean, the main regional environmental agreement is the 1983 Cartagena Convention and its three protocols which were developed under UNEP's regional seas programme (RSP).

- Cartagena addresses the management and conservation of the marine environment of the Caribbean and is complimented by three protocols: on oil spills, protected areas, and species and land-based sources of pollution.

- MEAs have had a tremendous impact on the law of individual states in the Caribbean region, and are a key tool in articulating environmental and natural resource policy and law at the national level.

∞ Links

See also 3.34 on the indigenous people of the Caribbean II.

Did you know?

Legend has it that the town was named Sauteurs, which is French for 'jumpers', because the last remaining Carib natives in Grenada jumped off a 40-metre-tall cliff later named Caribs' Leap to their deaths in 1651 rather than face domination by the conquering French. In Guyana, according to a Patamona Indian legend, Kaieteur Falls was named for Kai, a chief, or Toshao who acted to save his people by paddling over the falls in an act of self-sacrifice to Makonaima, the great spirit.

Did you know?

The Garifuna people, who are the descendants of Carib, Arawak and West African people, today live primarily in Central America, including Belize, but were 'exiled' from the island of St Vincent and the Grenadines to Roatán Island off the coast of Honduras in 1796.

The original peoples of the Caribbean region

The first inhabitants were nomadic tribes who migrated from Central America some 6,000 years ago. These indigenous people of the Caribbean have a long, rich, unique history and culture which is often overshadowed by the effects of European colonisation (Ferbel, 2000). The watershed event in the story of the indigenous peoples of the Caribbean appears to be the arrival of Columbus in 1492, and his subsequent naming of the West Indies. Before the arrival of Columbus, there were three groups of native peoples in the Caribbean: the Arawak, the Carib and the Ciboney. The Arawak populated the larger Caribbean islands of Cuba, Hispaniola, Jamaica and Puerto Rico. The Carib lived on the smaller volcanic islands of the eastern Caribbean: St Kitts-Nevis, Antigua, Guadeloupe, Dominica, Martinique, St Lucia, Barbados, St Vincent and Tobago. They had migrated earlier from the mainland of what is now called South America.

There are numerous accounts telling of how the indigenous people of the Caribbean were decimated by post-1492 colonial practices, including the interruption of agricultural scheduling, slavery, foreign disease, and genocide. Today, most of the islands of the Caribbean no longer have indigenous populations, but there remain in the history and culture in the form of folklore, archaeological remains, and place names such as the Kaieteur Falls in Guyana and Sauteurs in Grenada.

Indigenous rights are often the subject of protection under international instruments which place obligations on states to legally recognise, demarcate and effectively protect indigenous peoples' territories and natural resources. Guyana and Dominica are examples of Caribbean states which have made efforts to incorporate these principles at the national level.

Existing populations of indigenous peoples

Today there are five Caribbean countries that still contain populations of indigenous peoples – Belize, Dominica, Guyana, St Vincent and the Grenadines, and Suriname. Many of these people still maintain large parts of their traditional lifestyles and also have integrated into the fabric of their respective countries.

Belize has two main groups of indigenous peoples – those descended from the Mayans, and the Garifuna people, who were exiled to Roatán, an island off the coast of Honduras, by the British in 1796.

Dominica has an estimated population of about 3,000 Caribs, known as the Kalinago. They live in a remote and mountainous area of Dominica's Atlantic coast, on a 3,700-acre (15 km^2) Carib Territory or Carib Reserve which was established by the Carib Reserve Act, enacted the year of Dominica's independence in 1978.

Guyana has the largest remaining distinct tribes of indigenous people, or Amerindians as they are called. There are four main tribes, namely the Warraus, Arawaks, Wapisianas and the Caribs. The Caribs include several sub-tribes: the Arrecunas, Akawaios, Patamonas, Macusis, and the Wai-wais. The Warraus are believed by many scholars to be the oldest inhabitants of Guyana – this tribe is known archaeologically from the shell mounds of the North West and Pomeroon, some dating back 7,000 years. The Wai-wais are renowned as skilled weavers and bead workers and traditional Wai-wai architecture is exemplified in the Umana Yana (translated as 'Meeting Place') in Guyana's capital, considered by many Guyanese as the embodiment of Amerindian architecture.

St Vincent has two groups of Carib or Kalinago people – the 'Yellow Caribs' or Kalinago, and the 'Black Caribs' or the Garifuna. Garifuna are descended from a group of enslaved Africans who were marooned from shipwrecks of slave ships, as well as slaves who escaped to the island, who intermarried with the Carib and formed the last native culture to resist the British. In 1795 British colonists transported the Black Carib to Roatán Island, off Honduras, where they continue to live today. The Black Carib communities which remained in St Vincent are largely found in areas such as Sandy Bay and Oiwa, and still practise traditional activities such as whaling. The Caribs of Bequia are the only group of indigenous people in the Caribbean who are still allowed to whale under an indigenous exception to the 1946 Whaling Convention, despite an international moratorium on commercial whaling because of concerns about the extinction of many species of whales. The last known speakers of Carib in St Vincent died in the 1920s, and the language is now considered extinct.

The Amerindians, which are the original inhabitants of Suriname, form 3.7 per cent of the population. The main groups are the Akuriyo, Arawak, Carib/Kaliña, Trío (Tiriyó), and Wayana, and they live mainly in the districts of Paramaribo, Wanica, Maroni and Sipaliwini. Suriname, like Jamaica, also has Maroons, who are descendants of escaped West African slaves. Maroons make up 10 per cent of the population and are divided into five main groups: Ndyuka (Aucans), Kwinti, Matawai, Saramaccans and Paramaccans. Maroons have a similar lifestyle to the Amerindians, giving Suriname the distinction of having two distinct groups who practise traditional agriculture, fishing and forestry.

Figure 3.33.1 *The Maroons are descendants of escaped West African slaves*

Key points

- Indigenous inhabitants were the first settlers in the Caribbean region, and comprised three main groups: the Arawak, the Carib and the Ciboney.

- Because of European colonisation, most of the indigenous populations were either reduced drastically in size or annihilated, and today only five states have populations of indigenous peoples.

- Indigenous peoples were by their nature nomadic, and relied heavily on the environment for their food, shelter and basic necessities.

- They practice traditional agriculture, fishing and forestry, most of which is done at the subsistence level.

- Suriname and Jamaica have Maroons, who are descendants of West African slaves who escaped into the forest and highlands of their respective countries to avoid enslavement. Like the indigenous peoples, they also practise traditional agriculture, fishing and forestry, most of which is done at the subsistence level, and many of their populations exist today.

Learning outcomes

On completion of this section, you should be able to:

- identify sustainable resource management techniques used by indigenous peoples in the Caribbean region

- describe sustainable practices in fisheries and agriculture used by the indigenous peoples of the Caribbean region.

 Links

See also 3.33 on the indigenous people of the Caribbean I.

Did you know?

Slash-and-burn is also known as shifting cultivation, as indigenous peoples move around from one plot of land to another, allowing the forest to regenerate itself.

Did you know?

The definition of organic farming put forward by the International Federation of Organic Agriculture Movements (IFOAM), an international organisation for organic farming organisations established in 1972 is as follows:

Agriculture

Slash-and-burn

Indigenous people generally practise a nomadic lifestyle, which is based on the natural resources and features of the environment. One of the main features of their lifestyle is the growing of staple crops, especially cassava or manioc. The main form of agriculture is called slash-and-burn, where the vegetation of the area to be cropped is cut down, the remaining foliage set afire and the ashes used to provide nutrients to the soil for use of planting food crops. This is necessary, because the relatively shallow layer of humus common to tropical ecosystems means that the soil becomes infertile when the vegetative layer is removed.

The cleared area following slash-and-burn – known as swidden – is used for a relatively short period of time, and then left to fallow for a longer period – sometimes up to 10 or more years, to allow wild vegetation to grow on the plot of land. This form of cropping if practised properly can be sustainable, but as population densities increase, demand for land increases because of urbanisation and other commercial activities, and the fallow period by necessity declines. Many of the issues which contribute to the negative effects of slash-and-burn exist in the Caribbean.

Intercropping and organic farming

This is the growing of two or more crops simultaneously on the same piece of land. The practice promotes crop intensification in both time and space dimensions; it can yield greater quantities of crops as well as two or more crops per year. Intercropping may take various forms including mixed intercropping, row intercropping, strip intercropping and relay intercropping. Intercropping can maintain soil fertility as the different plant species, being of different sizes, take up nutrients from different layers of the soil; it can also reduce soil runoff, and produce a greater yield on a given piece of land by making use of resources that would otherwise not be utilised by a single crop. The practice of intercropping can also help limit outbreaks of crop pests by increasing predator biodiversity, because more than one plant species is present to serve as habitat for pest predators. Reducing the homogeneity of the crop increases the barriers against biological dispersal of pest organisms through the crop.

Organic farming is a form of agriculture that relies on techniques such as crop rotation, green manure, compost and biological pest control. Most traditional indigenous practices may be considered organic, since instead of artificial fertilisers, biological materials (such as manure and compost) and techniques (intercropping, shifting cultivation and crop rotation) are employed.

Organic agriculture is a production system that sustains the health of soils, ecosystems and people. It relies on ecological processes, biodiversity and cycles adapted to local conditions, rather than the use of inputs with adverse effects. Organic agriculture combines tradition, innovation and science to benefit the shared environment and promote fair relationships and a good quality of life for all involved.

Fishing

Fishing has played a rich and vital part in indigenous culture for centuries. In the Caribbean, fishing may include the capture of whales and cetaceans as well as fish. Beyond the importance of fishery products as a key food supply, techniques for catching them often bring indigenous populations together in a communal quest for food, and in touch with their spiritual practices. Indigenous peoples use a variety of techniques to fish – generally on a subsistence basis. Techniques include spear fishing, the use of blowguns – often laced with darts dipped in plant- or animal-based poisons such as curare or the poison dart frog (*Phyllobates* spp.), or other paralysing agents, the use of explosives and poisons to stun fish, fish traps and nets. Fishing may also be done from canoes, which indigenous peoples used to traverse waterways. Environmental reasons have forced governments in the Caribbean region to curtail and even prohibit some ancient techniques used by the indigenous peoples to catch fish, such as blast-fishing and the introduction of poisons into a river or stream to paralyse fish. However techniques such as spearing and trapping are still employed today. Also, because of the improvements in commercial fishing and the consumerist habits of man, many species relied upon by the indigenous peoples are threatened or in danger of extinction. Species include some marine turtles, whales and cetaceans, spiny lobsters (*Panulirus argus*), and the West Indian sea egg (*Tripneustes ventricosus*).

Use of forest: Timber and non-timber forest products

Indigenous forest people use their land in many different ways – for fishing, hunting, shifting agriculture, the gathering of wild forest products and other activities. For them, the forest is the very basis of survival and its resources have to be harvested in a sustainable manner.

While the forest is an important part of an indigenous lifestyle, in many countries the state is the official owner of most forest areas, even though some of the land may have been inhabited for generations by large numbers of people. Often the rights of the indigenous people are not recognised, and governments often need to balance the needs of the current inhabitants of the state with those of the original people.

Non-timber forest product (NTFP) is the term used for commodities obtained from the forest without the need for harvesting trees. NTFPs include game animals, nuts, oils, foliage and medicinal plants. Their economic, cultural and ecological value, when considered in aggregate, makes managing for NTFPs an important component of sustainable forest management and the conservation of biological and cultural diversity. In recent times, the value and use of these products has been highlighted and even promoted – as a means of maintaining the integrity of the forest ecosystem and the traditional values of the indigenous communities, as well as providing financial resources for these communities. This is the case of Dominica in the example given above in the Carib Reserve, and Guyana in the Iwokrama Reserve.

Key points

- Indigenous peoples were by their nature nomadic, and relied heavily on the environment for their food, shelter and basic necessities.

- They practise traditional agriculture, fishing and forestry, most of which is done at the subsistence level.

- Many of these traditional practices are considered sustainable, and have been promoted in pursuance of the goal of sustainable development.

Figure 3.34.1 *Indigenous Caribbean intercropping*

Did you know?

Curare is a common name for various arrow poisons used by the indigenous people of South America. Prey such as fish and small game is shot by arrows or blowgun darts dipped in curare, leading to asphyxiation owing to the inability of the victim's respiratory muscles to contract. Many countries have prohibited the use of these poisons for fishing, but they are still used in hunting game.

Did you know?

The practice of blast-fishing, which uses explosives to stun or kill schools of fish for easy collection, is no longer accepted, as it can be extremely destructive to the surrounding ecosystem.

Multiple-choice questions

1 The Kyoto Protocol is a sub-agreement of:
  A Convention on the International Trade in Endangered Species (CITES).
  B Convention on Biological Diversity.
  C United Nations Framework Convention on Climate Change.
  D United Nations Convention to Combat Desertification.

2 The Soufriere Marine Management Area (SMMA) is an area divided into five zones, located on the west coast of St Lucia. The SMMA was established to conserve the natural marine environment, and ensure sustainable use and development of the area, particularly within the fisheries and tourism sectors.
  Which term BEST describes this approach to resource management?
  A restoration
  B preservation
  C in-situ conservation
  D ex-situ conservation

3 Which of the following can be described as a renewable resource?
  A asphalt
  B limestone
  C gravel
  D seagrasses

4 The problems associated with withdrawing too much groundwater from a particular area are:
  I estuaries become saltier
  II the water table is lowered
  III land becomes less fertile
  A I and II only
  B I and III only
  C II and III only
  D I, II and III

5 Which of the following concepts is MOST CLOSELY associated with land use planning?
  A environmental impact assessments
  B zoning
  C user fees
  D recycling and re-use

Essay questions

1 a Define
    i ecotone [2 marks]
    ii wetland [2 marks]
  b Mangroves are a common type of vegetation found in wetland ecosystems.
    i Describe ONE function of mangrove ecosystems. [4 marks]
    ii Identify an international convention or agreement which may be used to manage and conserve wetland ecosystems. [2 marks]
  c The following results were obtained when sampling two sites in a wetland in Jamaica:

Species of mangrove and mangrove associates	Number of individuals	
	Site A	Site B
Avicenna germinans	55	6
Laguncularia racemosa	7	41
Conocarpus erectus	12	12
Rhizophora mangle	4	16

  Immediately after the study was carried out, a hurricane devastated both sites. Illustrate and explain which of the two sites is expected to recover faster from the disturbance. [6 marks]
  d Outline ONE measure which can be implemented to conserve mangrove ecosystems from anthropocentric activities. [4 marks]
  Total 20 marks

2 The sisserou (*Amazona imperialis*), which is the national bird of Dominica, is an endangered green-and-purple-plumaged Amazon parrot which is endemic to mountain forests of the island. It is featured on the Dominican flag, and one of the last remaining habitats of the sisserou (also known as the imperial Amazon) is on the slopes of Morne Diablotins, the highest volcanic peak of the Caribbean islands.
  a Define
    i Endemic [2 marks]
    ii Endangered [2 marks]
  b Outline THREE anthropogenic activities which may have an impact on the sisserou. [6 marks]
  c i Describe TWO measures which can be

implemented to manage and conserve this species. [8 marks]

ii Outline whether the sisserou being endangered and endemic would affect decisions regarding its management and conservation. [2 marks]

Total 20 marks

3 Figure 3.35.1 represents a traditional hunting practice utilised by indigenous peoples in the Caribbean.

Figure 3.35.1 *Traditional hunting practice*

a i Identify the traditional hunting practice identified in Figure 3.35.1. [2 marks]

ii Outline how this type of hunting practice is carried out. [4 marks]

iii In many Caribbean islands, this traditional hunting practice is no longer permitted. Identify TWO reasons for this. [4 marks]

b A gold company plans to exploit a gold reserve in a remote part of a Caribbean state. The location of the reserve is heavily forested and hilly, with few population centres.

i Outline TWO factors which the government of the Caribbean country may have considered before making the decision to grant permission to exploit the reserve. [8 marks]

ii Identify one tool of natural resource management which the government may require before the company is allowed to exploit the reserve. [2 marks]

Total 20 marks

4 Figure 3.35.2 shows the effect of increased fishing on the fishery resources of a Caribbean island.

a i Describe the trend observed in Figure 3.83.2. [3 marks]

ii Using the graph, determine the fishing effort which will produce a daily catch of 2500 kg of fish. [2 marks]

b i What is meant by the term 'maximum sustainable yield' in relation to the harvesting of fish stock? [2 marks]

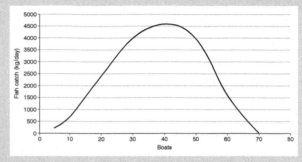

Figure 3.35.2 *Effect of fish catch on harvesting effort*

ii At which of the three points, A, B or C, shown on Figure 3.35.2 should harvesting take place to achieve the maximum sustainable yield? [1 mark]

iii Give ONE reason why it is recommended to harvest a fishery resource at the maximum sustainable yield. [3 marks]

c Explain the effect of EACH of the following on the harvesting of the fishery resource in Figure 3.35.2:

i Legislation

ii Coastal zone management

iii Community-based natural resource management. [8 marks]

d Identify ONE conservation measure which may be used to improve the status of the fishing resource. [1 mark]

Total 20 marks

1 Presenting and interpreting data for the Internal Assessment I

The Internal Assessment

The Internal Assessment is an integral and compulsory part of student assessment for CAPE Environmental Science. It is intended to assist students in acquiring certain knowledge, skills and attitudes that are associated with the subject. The Internal Assessment should relate to at least ONE specific objective in the unit. It must always be remembered that the activities for the Internal Assessment should be linked to the syllabus and should form part of the learning activities to enable students to achieve the objectives of the syllabus.

It is important to note that when planning for the Internal Assessment it should be conceived as a single research project which would allow for the three components to be realised and then integrated. This approach would allow for adequate linkage and logical consistency among the three components: site visits, laboratory exercises and final report.

The reports for the series of site visits and associated laboratory exercises should be recorded in the journal which should comprise:

a   an entry for each site visit

b   a report for the journal

c   a final report on the set of site visits.

Each student is expected to conduct and write a final report on a **minimum** of four (4) site visits and four (4) laboratory exercises.

Components of the Internal Assessment

Site visits

Site visits and activities chosen for site visits could have either a spatial or cross-sectional **OR** a temporal or longitudinal element. In this regard activities should be based either on visits to **one** site where changes over a period of time are observed **OR** on a series of visits to different sites to compare and contrast similar processes or occurrences.

The entry for each site visit should be recorded using the format below:

i   Entry Number

ii   Date

iii   Site (Location)

iv   Objective(s)

v   Activities

vi   Observations

vii   Comments

viii Follow-up activities.

Laboratory activity and laboratory report

A laboratory report for each of at least four (4) laboratory exercises is expected. Laboratory exercises should relate to each or any of the series of site visits.

The areas that will be assessed in the report for each **laboratory exercise** are:

a Planning and Designing;

b Observation and Recording;

c Manipulation and Measurement;

d Analysis and Interpretation;

e Reporting and Presentation.

EACH laboratory exercise should be reported using the format below:

i Title

ii Aim

iii Materials

iv Procedure

v Data Collection/Results

vi Discussion and Conclusions.

A note of caution

Avoid conducting laboratory investigations for the same element or condition at the four site visits since such an approach will constitute only one (1) laboratory exercise. If, for example, water is tested at each of the four different site visits for pH, the four tests for pH will be recognised as one (1) laboratory exercise (investigating one parameter) but done for four samples. It is also advisable that the same elements or conditions are investigated at each of the four site visits. This allows for comparisons and better explanation of phenomena among sites and during different site visits.

The final report

The entries for the site visits and the reports for the laboratory exercises **MUST** inform the final report for the journal. The final report must not exceed **1500** words. There is a penalty for exceeding the specified word limit. The areas that are assessed in the final report for the journal are summarised below.

1 Project description (clarity of the statement of the problem being studied)

2 Purpose of the project (definition of the scope of the project)

3 Adequacy of information/data gathered and the appropriateness of the design chosen for investigating the problem

4 Appropriateness of the literature review

5 Presentation of data/Analysis of data (summary of site visits and laboratory exercises)

6 Discussion of findings

7 Conclusion

8 Recommendations

9 Communication of information

10 Bibliography.

Key points

- The Internal Assessment is an integral and compulsory part of student assessment for CAPE Environmental Science and should relate to at least ONE specific objective of the syllabus.

- The Internal Assessment should be conceived as a single research project which would allow for the three components: site visits, laboratory exercises and final report to be linked.

- Site visits and activities chosen for site visits could have either a spatial or cross-sectional **OR** a temporal or longitudinal element.

- It is advisable that the same elements or conditions are investigated at each of the four site visits. This would allow for comparisons and better explanation of phenomena.

- Entries for the site visits and the reports for the laboratory exercises **MUST** inform the final report for the journal. The final report must **NOT EXCEED 1500 words**.

Learning outcomes

On completion of this section, you should be able to understand how to:

■ complete a journal entry

■ interpret the results obtained and observations made during the site visit

■ apply the skills that are to be assessed

■ link the objectives of the components of the Internal Assessment.

Site visits and journal entry

Each student is required to complete a journal in which certain specific practical skills should be demonstrated. The record of the site visit must include the following:

Entry number	Indicates the position and the number of the entry.
Date	The actual date the site was visited.
Site (Location)	A BRIEF description of the location or direction on how to get to the location. A diagram/map is usually very useful.
Objective(s)	There should be a relevant objective or a set of relevant objectives for each site. Objectives should give a clear indication of what is to be achieved and how it will be achieved.
Activities	The series of events undertaken at the site. Should be systematic and give details.
Observations	Observations should focus on the objectives and students should make observations of everything at the site. Observations should not be limited solely to results from tests for parameters. Always observe and make notes on the surrounding environment.
Interpretative comments	This is an opportunity to provide a scientific explanation for the results obtained and the observations made. It must be factual and conclusive and based on an interpretation of the observations and results.
Follow-up activities	This is an opportunity to say what will be done for the next site visit. It can indicate when the next site visit will take place, the location of the visit, what observations and investigations will be done and how the results, data and information will be used.

Example of journal entry record

Title: A comparison of the abundance and percentage frequency of three weed plant species in two school gardens: the CAPE school garden and the CXC school garden.

Entry number	1
Date	10 May 2013
Site (Location)	CAPE School garden, Longman Road, Caribbean
Objective(s)	To compare, by means of the quadrat method, the abundance and percentage frequency of three weed plant species in the CAPE school garden and the CXC school garden.
Activities	■ Make observations of the different weed species in the CAPE school garden. ■ Identify the locations where quadrats will be placed for taking measurements. ■ Finalise the number of quadrat throws that will be made.

	■ Place quadrats and record information on the three weed plant species. ■ Make observations of the activities that take place in surrounding areas and in close proximity to the CAPE school garden.
Observations	■ Students were working in other areas of the CAPE school garden. ■ There was a pond in the garden that was almost overgrown with aquatic weeds. ■ A dense patch of bamboo vegetation was observed growing on the periphery of the garden. ■ Lots of different vegetables were harvested by students and being sold to the teachers.
Interpretative Comments	■ Vegetables appeared to be the main crops grown on the CAPE school garden. ■ The lush growth of aquatic vegetation in the pond may have been supported by fertiliser runoff from cultivation into the pond. ■ The dense vegetation of the bamboo trees provided shade for the cultivated crops in the garden.
Follow-up activities	■ Analyse the data collected for the three weed plant species and calculate the percentage frequency for each of the three species. ■ Plan the date and time for the visit to the CXC school garden to collect data on species abundance and percentage frequency for the species observed.

Sample calculation that could be done

Quadrat Sampling Data: A table is useful for summarising quadrat data.

Formula for calculating percentage frequency

$$\text{Percentage frequency} = \frac{\text{number of quadrats with species}}{\text{total number of quadrats}} \times 100$$

Relative abundance of species

Species	Mucuna pruriens	Cordia curassavica	Commelina cayennensis
Quadrat 1	10	15	5
Quadrat 2	11	0	0
Quadrat 3	15	20	0
TOTAL	36	45	45

Calculations:

Percentage frequency (*Mucuna pruriens*) $= \dfrac{3 \times 100}{3} = 100\%$

Percentage frequency (*Cordia curassavica*) $= \dfrac{2 \times 100}{3} = 66.67\%$

Percentage frequency (*Commelina cayennensis*) $= \dfrac{1 \times 100}{3} = 33.33\%$

Did you know?

- Data is processed to determine what patterns and relationships emerge.
- Data analysis provides opportunities for students to identify values that are inconsistent and for comparisons to be made with local, regional and international standards.
- The use of more than one analytical technique is encouraged. Available techniques that could be used include: averages, percentages, mean, mode, median, variance, standard deviation.
- The use of different ways of presenting the data is also encouraged.

Key points

- Include a BRIEF description of the location or direction on how to get to the location. A diagram or map is usually very useful.
- There should be a relevant objective or a set of relevant objectives for each site. Objectives should give a clear indication of what is to be achieved and how it will be achieved.
- Interpretative comments must be factual and conclusive and based on an interpretation of the observations and results.
- Follow-up activities MUST always be included in the journal entry. This is an opportunity to say what will be done for the next site visit, when the next site visit will take place, the location of the visit and the objectives of the next site visit.

3

Presenting and interpreting data for the Internal Assessment III

Learning outcomes

On completion of this section, you should be able to understand how to:

- plan and design a laboratory exercise

- present and interpret the results obtained

- apply the skills that are to be assessed

- link the objectives of the components of the Modules in the respective Units for the Internal Assessment.

⚭ Links

See the CD for an example of a laboratory activity and record.

☑ Exam tip

When analysing data, it is important to identify the overall trends shown; this may in fact be a common question. Note the highs, lows and other significant points from the data presented. If the rate of growth or decline is significant, or changes over time, this should be noted. If there is a clear relationship between the data presented, identify the nature of this relationship.

Did you know?

Discrete sites could be visited over time for taking specific measurements. In this way students can collect and analyse data that demonstrates change over time for a particular phenomenon or factor at a specific location.

The laboratory exercise

Planning and designing	This entails pre-planning by the students and teacher to determine and set objectives for the activity and also to finalise what parameters will be investigated and how these investigations will be conducted.
Observation and recording	The first step of the scientific method involves making an observation about something that is of interest and recording what is observed.
Manipulation and measurement	Students are expected to be able to use different instruments to measure various parameters accurately and efficiently. They should also know the basic working principles of the instruments.
Analysis and interpretation	Analysis of data gives order and meaning to the data collected. It facilitates interpretation and comparison of different types of data to determine trends. Interpretations must be factual and based on the data presented and analysed.
Reporting and presentation of results	A standard format for writing a laboratory report exists and should be followed. The results are where you report what happened in the experiment, detailing all observations and data collection carried out during your experiment. Data presentation involves the description of data or results obtained. It is often easier to depict the data in visual form by charting or graphing the information. Different techniques of data presentation could be used, including tables, graphs, maps, diagrams, sketches and photographs.
Methodology/ Procedure	Methodology and procedure refers to the set of procedures for all test kits used. It also describes the actual procedure that is followed for sample collection, making observations, and testing of samples.
Conclusion	The final step of the scientific method is the conclusion. This is where all of the results from the experiment are analysed and a determination is reached about the hypothesis. Did the experiment support or reject your hypothesis? If your hypothesis was supported, great. If not, repeat the experiment or think of ways to improve your procedure.

Data collection/results:

Presentation of data: Table

Results for measurements taken at six (6) different sampling stations

	Sampling Station A	Sampling Station B	Sampling Station C	Sampling Station D	Sampling Station E	Sampling Station F
Temperature	27.8 °C	27.9 °C	28.9 °C	28.8 °C	29.2 °C	29.2 °C
pH	8	9	10	10	9	10
Turbidity	3 NTU	4 NTU	5 NTU	5 NTU	4 NTU	4 NTU

Presentation of data: Line graph

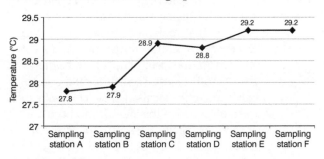

Figure 4.3.1 *Temperature of river at different sampling stations*

Presentation of data: Bar chart

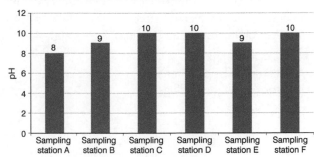

Figure 4.3.2 *pH of river at different sampling stations*

Presentation of data: Bar chart

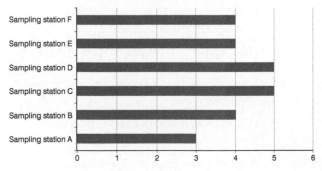

Figure 4.3.3 *Turbidity of river measured at different sampling stations. Measurements are in nephelometric turbidity units (NTU)*

Key points

- A standard format for writing a laboratory report exists and should be used.

- The planning and design phase determines and sets objectives, and finalises the parameters and methodology of investigation.

- Students are expected to be able to use different instruments. They should therefore know the basic working principles of the instruments.

- Analysis of data is important since it gives order and meaning to the data collected and facilitates interpretation and comparison of data to determine trends. Interpretations must be factual and based on the data presented and analysed.

- The final step of the scientific method is the conclusion.

Did you know?

The Internal Assessment component is compulsory and the skills that are assessed are:

- selection of techniques, designs, methodologies and instruments appropriate to different environmental situations;

- collection and collation of data;

- analysis, interpretation and presentation of such data;

- use of appropriate quantitative techniques;

- development of appropriate models as possible solutions to specific environmental problems.

Presenting and interpreting data for the Internal Assessment IV

On completion of this section, you should be able to understand how to:

- prepare the final report of the investigation

- apply the skills that are to be assessed

- draw appropriate conclusions based on the results and findings

- make appropriate recommendations that are informed by the findings and conclusions.

The final report

The final report is informed by the entries for the site visits and the reports for the laboratory exercises. The final report **must not exceed 1500** words. The areas that are assessed in the final report for the journal are summarised below:

Final report for journal	Descriptors and expectations
Clarity of the statement of the real-world problem being studied (project description)	The problem statement could be a brief introductory paragraph that describes the problem. The problem statement must be clear and concise. The title chosen must be specific and concise and ideally should not be more than 12 to 15 words. The title should give an indication of the location(s) being investigated, the time period of the investigation and the phenomenon being observed. The research problem must relate to one or more modules within the specific unit of the CAPE Syllabus.
Definition of the scope of the project (purpose of project)	The purpose should be stated and all variables must be identified. As part of this process the title may be restated and elaborated. After reading the purpose or scope one should get an indication of how the problem being investigated will be addressed.
Methodology	The way in which the project was undertaken should be summarised and assessed here. Research design and justification of this design/methodology. This includes methods used both in the site visits and the practical exercises.
Appropriateness of the literature review	There must be evidence that an appropriate and comprehensive literature review was conducted. The literature review is actually a general background and introduction to the problem being investigated and it may also present a summary of related work similar to that being investigated. A good literature review can also help to inform the methodology chosen, the variables identified and the techniques that could be used for measuring them. A good literature review also illustrates what other researchers would have done on similar research, identifies gaps in the literature, suggests areas for future and further research and identifies recommendations made by previous researchers.

	Overall, a comprehensive literature review provides an opportunity to inform about the topic being investigated.
Presentation of data/ Analysis of data	The final report must demonstrate evidence of adequacy of information, data gathered and the appropriateness of the design chosen for investigating the problem.
	Data collected could be presented using various graphs, tables, figures and statistical symbols, maps, diagrams and photographs, creatively and adequately.
	There must be evidence of data analysis that is adequate and which utilises **at least** two or more approaches. Good data analysis reveals trends, patterns and relationships and allows for effective evaluation and identification of findings.
Discussion of findings	All findings must be clearly stated, supported by data and their interpretability addressed. The reliability, validity and usefulness of all findings must be addressed.
Conclusion	Conclusions made must be clear, concise, based on finding(s), valid and related to the purpose(s) of the project.
Recommendations	Recommendations made must be fully derived from the findings.
Communication of information	Information should be communicated in a logical manner with no grammatical errors.
Bibliography	References must be written using a consistent convention and there is a minimum requirement of four references.

Key points

- The final report must be informed by the entries for the site visits and the reports for the laboratory exercises.

- The final report **must not exceed 1500** words. There is a penalty for exceeding the word limit.

- The problem statement must be clear and concise and the title must be specific and concise.

- The final report must demonstrate evidence of a comprehensive and relevant literature review, adequacy of information, data gathered and the appropriateness of the design chosen for investigating the problem.

- All findings must be clearly stated, supported by data and their interpretability addressed.

- The reliability, validity and usefulness of all findings must also be addressed.

- Clear and concise conclusions must be made and these must be based on finding(s), be valid and be related to the purpose(s) of the project.

- Recommendations must only be derived from the findings.

References

Beale, M. (2010). Management of the Blue and John Crow Mountains National Park by JCDT/Green Jamaica. In: Forests for people, people for forests: forest-based livelihoods in the Caribbean. Port-of-Spain, Trinidad: CANARI. 4–7 May 2010.

Borrini-Feyerabend, G., Kothari, A. and Oviedo, G. (2004). *Indigenous and local communities and protected areas: towards equity and enhanced conservation.* Gland, Switzerland and Cambridge, UK: IUCN. ISBN 2831706750.

Borrini-Feyerabend, G., Pimbert, M., Farvar, M.T., Kothari, A. and Renard, Y. (2004). *Sharing power: learning by doing in co-management of natural resources throughout the world.* Tehran, Iran: IIED and IUCN/CEESP/CMWG, CENESTA).

Brown, N.A. and Bennett, N.G. (2010). Consolidating change: lessons from a decade of experience in mainstreaming local forest management in Jamaica. Laventille, Trinidad: CANARI.

Callicott, B. (1989). *In defense of the land ethic: essays in environmental philosophy.* Albany, NY: State University Press of New York Press.

Ferbel, P. J. (2000). Book Review: *The indigenous people of the Caribbean.* Ed. Wilson, S.M. Gainesville, University Press of Florida, 1997. *Ethnohistory* , *47.3–4*, 816–818.

Government of Grenada, Ministry of Tourism (2006). Report of Conference on Community Tourism Development In Grenada.

Guarderas, A.P., Hacker, S.D., Lubchenco, J. (2011). Ecological effects of marine reserves in Latin America and the Caribbean. *Mar. Ecol. Prog. Ser.* 429: 219–225.

Guarderas, A.P., Hacker, S.D., and Lubchenco, J. (2008). Current status of marine protected areas in Latin America and the Caribbean. *Conservation Biology, 22,* 1630–1640.

Heileman, S. and Walling, L.J. (eds.) (2005). *Caribbean environment outlook.* UNEP/Earthprint. ISBN 9789280725261.

Hinrichsen, D. (1998). Coastal waters of the world: trends, threats, and strategies. Washington D.C.: Island Press. Online at: http://oceanservice.noaa.gov

McDermott, M. (2010). Improving community livelihoods though co-management of the watershed: a case study of the Fondes Amandes Community Reforestation Project. Laventille, Trinidad: CANARI.

Pantin, D., Reid, V. and Maurice, N. (2006). Valuation of the contribution of the Talvern Water Catchment Group, Saint Lucia, to watershed management. Who pays for water. Project Document No. 2. Laventille, Trinidad: CANARI.

Rolston, H. (1986). *III. Philosophy gone wild: essays in environmental ethics.* Amherst, NY: Prometheus.

Soulé, M. E. (1985). What is conservation biology? *Bioscience, 35,* 727–734.

Stanford, J.A., Ward, J.V., Liss, W.J., Frissell, C.A., Williams, R.N., Lichatowich, J.A. and Coutant, C.C. (1996). A general protocol for restoration of regulated rivers. *Regulated Rivers: Research & Management, 12,* 391–413. Online at: http://digitalcommons.unl.edu

Trewenack, G. (2010). Conserving the Grande Riviere watershed: a case study of collaborative forest management in north-east Trinidad. Laventille, Trinidad: CANARI.

Index